Research towards the Green and Sustainable Buildings and Cities

Research towards the Green and Sustainable Buildings and Cities

Guest Editors

Haibo Guo

Bolun Zhao

MDPI

Basel • Beijing • Wuhan • Barcelona • Belgrade • Novi Sad • Cluj • Manchester

Guest Editors

Haibo Guo
School of Architecture and Design
Harbin Institute of Technology
Harbin
China

Bolun Zhao
School of Architecture and Design
Harbin Institute of Technology
Harbin
China

Editorial Office
MDPI AG
Grosspeteranlage 5
4052 Basel, Switzerland

This is a reprint of the Special Issue, published open access by the journal *Buildings* (ISSN 2075-5309), freely accessible at: https://www.mdpi.com/journal/buildings/special_issues/5B1779A081.

For citation purposes, cite each article independently as indicated on the article page online and as indicated below:

Lastname, A.A.; Lastname, B.B. Article Title. *Journal Name* **Year**, *Volume Number*, Page Range.

ISBN 978-3-7258-5157-7 (Hbk)
ISBN 978-3-7258-5158-4 (PDF)
https://doi.org/10.3390/books978-3-7258-5158-4

Contents

About the Editors

Haibo Guo

Haibo Guo is a Professor at the School of Architecture and Design, Harbin Institute of Technology (HIT), in Harbin, China. His research focuses on green building materials, building energy efficiency, and sustainable urban development, aiming to advance innovative solutions for environmentally responsible construction and urban planning. He has led or participated in 10 major national-level research projects, including grants from the National Natural Science Foundation of China (NSFC), addressing critical challenges in sustainable building and city design. His work has contributed significantly to developing practical frameworks for enhancing the ecological performance of built environments. Professor Guo actively supports academic discourse through his roles as a Guest Editor and peer reviewer for several leading international journals in the fields of sustainable architecture and urban science.

Bolun Zhao

Bolun Zhao is a doctoral candidate at the School of Architecture and Design, Harbin Institute of Technology (HIT), in Harbin, China. His research focuses on building indoor thermal comfort, computational design methodologies, and sustainable biomass materials for construction applications, with the overarching goal of advancing sustainable development in buildings and cities. As an active researcher, he has authored or co-authored more than 20 peer-reviewed publications contributing to these fields. Mr. Zhao further supports the academic community by serving as a reviewer for several leading peer-reviewed journals in architecture, building science, and sustainable materials. His work consistently bridges theoretical innovation with practical solutions for enhancing environmental performance in the built environment.

Preface

Our present era demands a radical rethinking of the built environment. Climate change, resource depletion, greenhouse gas emissions, and sustainable development imperatives compel us to embrace energy-efficient, ecologically sensitive, and sustainable practices. This Reprint, Research towards the Green and Sustainable Buildings and Cities, responds to this critical need.

The scope of this collection spans the transformative potential of emerging technologies—such as green building techniques, artificial intelligence, and digital twins—alongside the vital preservation of architectural heritage and cultural identity. It explores how innovation harmonizes with tradition to revitalize communities and redefine sustainable architecture.

The aim of this Reprint is to showcase cutting-edge research driving multidisciplinary solutions for sustainable urban futures. We are motivated by the urgency of global challenges and the opportunity to steer architectural practice toward resilience, cultural continuity, and ecological stewardship.

This Reprint presents 18 peer-reviewed papers advancing knowledge in green materials, smart building systems, heritage-sensitive retrofits, and urban sustainability frameworks. We extend our deepest gratitude to the contributing authors for sharing their expertise, the diligent reviewers for ensuring scholarly rigor, and the Buildings editorial team for their unwavering support.

We address this work to researchers, architects, urban planners, engineers, policymakers, and all stakeholders committed to forging sustainable, livable cities and buildings for generations to come.

Haibo Guo and Bolun Zhao
Guest Editors

buildings

MDPI

Article

A Rule-Based Design Approach to Generate Mass Housing in Rural Areas of the North China Plain

Jiang Wang [1,*], Wei Fan [2], Bolun Zhao [3,4], Yang Yang [1] and Zhijun Zhang [5]

[1] School of Architecture and Urban Planning, Shandong Jianzhu University, Jinan 250101, China; 24265@sdjzu.edu.cn
[2] School of Architecture, Tianjin University, Tianjin 300354, China; 1022206003@tju.edu.cn
[3] School of Architecture, Harbin Institute of Technology, Harbin 150001, China; 22b934023@stu.hit.edu.cn
[4] Key Laboratory of Cold Region Urban and Rural Human Settlement Environment Science and Technology, Ministry of Industry and Information Technology, Harbin 150001, China
[5] School of Computer Science and Technology, Shandong Jianzhu University, Jinan 250101, China; 12778@sdjzu.edu.cn
* Correspondence: wangjiang@sdjzu.edu.cn

Abstract: Affected by the development strategy of Rural Space Reconstruction in China, the demand for rural mass housing has peaked in the North China Plain in the past 20 years. However, due to the inefficiency of conventional design methods, the rural houses built appear to have a noticeable trend of urbanization and homogeneity. To propose a more effective design approach to change the hitherto unsuccessful homogenized phenomenon of rural design, the study is based on investigating the composition, configuration and characteristics of the dwellings in some traditional villages of the eastern Shandong Province, and it compares and analyzes the differences between conventional methods and generation methods through three design tests: Test 1 is for the reappearance of a general mode of planning, Test 2 is based on the definition of shape-grammar-based rules and Test 3 is mainly used for the optimizing and programming of rules. Furthermore, based on the three prototypes of homestead combination, three-level rules are determined through the three tests mentioned above: Level-1 describes the housing prototype consisting of four homesteads, which generates a variety of spatial relations through the translation of homesteads. Level-2 describes a neighborhood prototype consisting of 16 homesteads, which generates various samples through splitting prototypes and expanding homesteads. Level-3 describes a block prototype consisting of 64 homesteads, which controls open space and identifies a given base during sample filling. Through the analysis of the tests results, the rationality and feasibility of the generative design approach are verified, proving that this approach effectively solves the design monotony problem that commonly exists in rural mass housing in the North China Plain.

Keywords: rule-based approach; generative design; mass housing; rural housing; shape grammar; design process; automation; space syntax

1. Introduction

Since 2000, the Chinese government has implemented the development strategy of Rural Space Reconstruction. Through planning and control, the government has led and planned rural settlements and promoted the moderate concentration of the rural population. On this basis, it is necessary to allocate the rural infrastructure appropriately, improve the quality of human settlements in rural areas, and form a spatial structure conducive to urban–rural integration [1]. To make more compact and efficient use of rural land resources in the North China Plain, some local governments have mandated that architects design mass rural housing, which may take the form of either urban multi-story houses with upstairs living spaces or single-door, single-courtyard row houses. However, upon the completion of certain projects, the homogenized and dull image clashed with the original rural aesthetic, leading to scrutiny from many architectural critics. This unified construction of mass rural housing appears to be evolving into a newly emerging architectural typology, thereby impacting the rural construction process in China. Because this type of building belongs to rural projects, its low price makes it difficult for architects to devote a lot of energy, and many projects are designed using simple row houses. Furthermore, in the past, rural houses were typically constructed by individual families, leading to diverse and evolving spatial relationships among households through diachronic evolution. In a unified construction cycle, if the traditional spatial relationships are to be reconstructed simultaneously, the overall planning of these small body and large quantity houses should align with the fundamental elements of traditional styles. This calls for the establishment of a set of efficient design methods, abandoning the conventional blueprint-based design approach.

With this scenario in the backdrop, the current study employed a rule-based approach to the design of rural housings and homesteads to generate a variety of layout patterns that represented the traditional Chinese rural settlement. The results of this study showed predefined, phased, hierarchical, and differentiated characteristics. The findings are expected to provide a foundation for the industrialized methods of construction which may further support automated design and mass customization.

2. Literature Review

With the advent of digital design tools, designers' awareness and handling of complex spatial issues improved significantly. Not only can a digital design produce a personalized response to real-world design issues, but it can also provide a meaningful alternative to repetitive design logic. Based on Gilles Deleuze's philosophical theory, Oxman [2] summarized five digital design models, viz. the CAD model, digital formation model, generative design model (GDM), performance model, and compound model. Among these models, the GDM is used to define the calculation rules of the formal generative mechanism, and the results obtained are more flexible and can maximize the implementation of interactional design. It has been widely used in the field of digital architecture design recently.

The GDM framework consists of several different design approaches or models, such as cellular automata (CA), genetic algorithms (GA), L-systems (LS), shape grammars (SG), and swarm intelligence (SI). Most of them, like SG, are branches of the rule-based system, which is one of the most basic automation design techniques. Although there are some similarities and overlaps between them, each approach deals with different design issues [3]. For instance, SG is generally used when there is a strong domain of knowledge [4], such as in the study of mass housing layout.

SG can be both descriptive and generative. It can be used both as an analytical tool to describe the formal structure of a corpus of existing designs, and as a synthetic tool to create new languages of design. Initially, SG was used for the generative specification of painting and sculpture [5]. Stiny and Mitchell [6] defined SG that constituted the style of Palladio's villas and translated them into concrete methods and a generative process. Chiou and Krishnamurti [7] were the first to use SG to study folk houses in Taiwan, which established the fact that SG can be applied to study Chinese traditional architecture. Duarte [8] described the grammar of Alvaro Siza's houses at Malagueira, and applied an interactive computer system to design customized mass housing. In recent years, several computational design methods have been applied on a large scale. Beirão et al. [9] defined Urban Grammar formally and developed a digital design tool to start the research of a customized city. Therefore, SG is being increasingly used to clarify any design rules of complex architectural designs and urban planning, leading to the building of a multi-level design corpus [10].

Considering Duarte [11] and Wang et al. [12], the residential design world can be described as a hierarchical structure comprising materials, components, parts, rooms, dwellings, housings, neighborhoods, blocks, settlements, and towns, of which the housing, neighborhoods, and blocks have been the focus of the present study (Figure 1). Regarding housing, the individual buildings which comprised similarly shaped elements [13] were analyzed, such as in the case of Erem et al. [14] who added Turkish rural housing grammars to the new rules, which can improve the adaptability of traditional housing after reconstruction. With regard to neighborhood, combinations of similar housing were considered, such as in the case of Herbert et al. [15]. With respect to blocks, mainly the roads, layout, and scale were paid attention to, such as in the case of Dias [16].

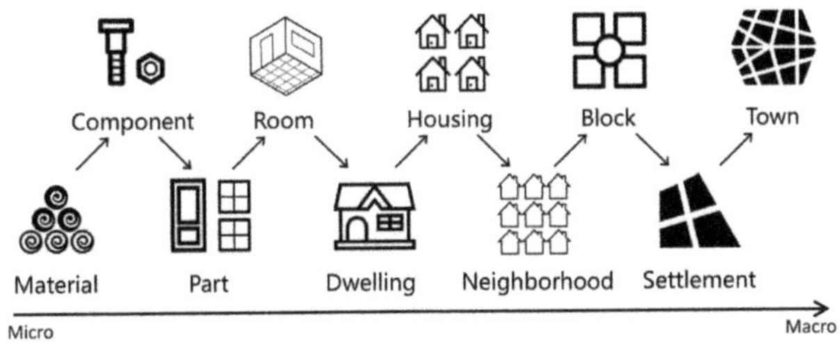

Figure 1. Hierarchical structure of mass customized housing design.

3. Methodology

To propose a more effective method, selecting an actual base for design research is necessary. Therefore, Zhuo village in the North China Plain was selected as the target, and an investigation was conducted on it and some surrounding villages to clarify the composition, configuration and characteristics of typical rural housing in the local area. And, then, we qualitatively proposed a housing prototype, core housing, and their relative relationship with homesteads. In the subsequent research on the rule-based approach, on the one hand, some design tests were conducted using the base of Zhuo village. On the other hand, a series of applicable rules have been proposed, some of which come from the investigation results and some from the professional experience of the researchers.

3.1. Investigation and Analysis Approach

The first method was a field investigation. In this method, 16 villages were selected which are distributed in the plain areas of the Wulong River, adjacent to Laiyang City of the Jiaodong Peninsula (Figure 2). These villages are all urban fringe villages with a history of more than 200 years and comprising 100 to 2000 households. Due to the rapid surge in demand for urban entertainment and rural tourism in recent years, these villages have gradually transformed their traditional farming to experiential agriculture. In our initial investigation, Google Earth images were used to analyze the boundary, road, layout and scale of each village. Next, through field research, particularly the texture of the Zhuo village was focused to systematically analyze the three levels—block, neighborhood and housing. At the block level, the roads were divided into main roads with a width of about 6 m and alleys with a width of about 3–4 m. The main road is used for household and agricultural vehicles, and the alleys are primarily used for walking. Tortuosity and cul-de-sacs are the two basic forms of the roads; the homesteads adjacent to the main road are larger because they are often used for commercial services. With regard to the neighborhood aspect, most of the blood-related families were found to live next to each other. This is due to the influence of the traditional family culture. Later, as the economy developed and population mobility increased, the closely related families split into single-family households with a small courtyard. At the housing level, five typical fragments consisting of four detached houses were selected for diagram and transformation to designate a housing-level prototype (Figure 3). In addition, the core housing faced south with a maximum of 2 floors and rooms having widths between 3 and 3.3 m. The shape of the homestead mainly included two types: long rectangle in a vertical direction and approximate square. Furthermore, the gable of the core housing and the northern boundary of the homestead were usually not provided with any gates.

Figure 2. The distribution map of villages investigated. ① Bozhuang village. ② Yuanshang village; ③ Heshankou village. ④ Dongzhuang village. ⑤ Dongzhuanggezhuang village. ⑥ Beisonggezhuang village. ⑦ Dazhuangzi village. ⑧ Nanligezhuang village. ⑨ Beihuang village. ⑩ Xuzhuangzhuang village. ⑪ Fengtou village. ⑫ Donggezhuang village. ⑬ Heqian village. ⑭ Cuizhuang village. ⑮ Jiangzhuang town. ⑯ Zhuo village.

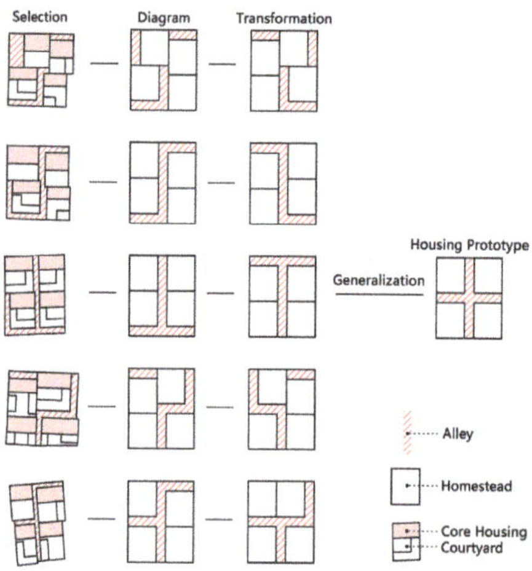

Figure 3. The evolution map of housing prototype.

3.2. Rule-Based Approach

The second method was the automation of rule-based design, which was divided into three stages (Test-1, 2, 3) of progressive implementation. Specifically, the second method included the definition of rules, implementation of rules, comprehensive analysis and the encoding of rules. (1) In the field investigation, we used SG-based rules to derive the different housing prototypes. We also defined the rules for non-SG (non-Euclidean transformation) at all the 3 levels—the housing, neighborhood and block levels (homesteads from 4 to 16, 64). Moreover, the combination rules and filling rules were also defined. (2) Both qualitative and quantitative analyses were performed to verify the results of every test. In the qualitative analysis, the results of the tests were compared with traditional settlements, which showed whether the test results had the characteristic form and organization of a traditional residential space. In the quantitative analysis, Space Syntax was applied. The Space Syntax can be used to quantitatively analyze the spatial relationships implied in environmental design, interpret the impact of urban form and land use for pedestrians and bicycle transportation, and assist in decision-making in the planning and design of future environments. The integration value of the tests results was measured, and the high integration value represented high reachability and great potential to become a center [17]. (3) Furthermore, using the preliminary results, the filtration rules, classification rules, and adjustment rules were set to optimize the whole design process. The rules were coded via algorithms using Python and applied in a computer program to develop the design corpora. Subsequently, we selected a real base from the Zhuo village which needs to be renewed and used Pyautocad to identify its boundaries and roads. Samples from the corpora were then randomly and automatically filled into the base following the developed rules in order to realize design automation (Figure 4).

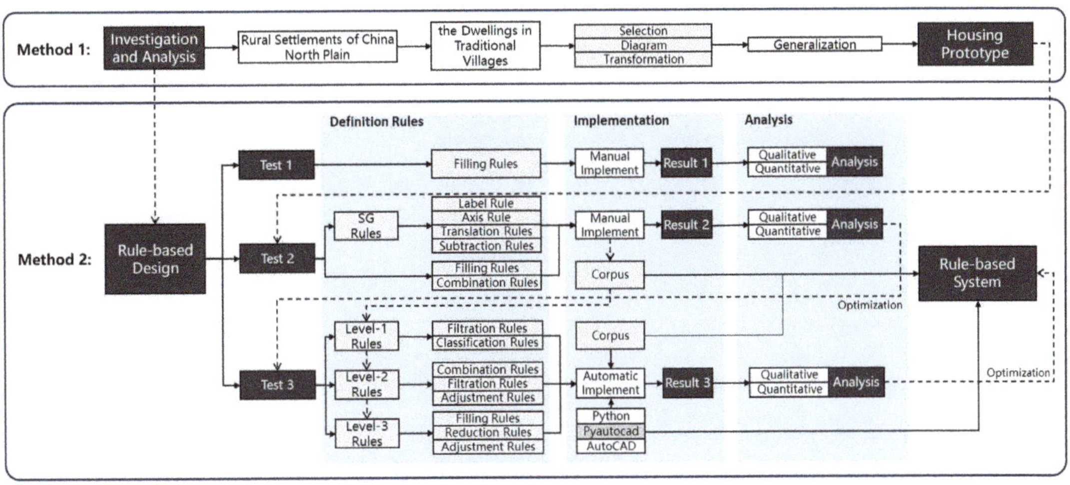

Figure 4. Process of rule-based design framework.

4. Predefinition

A basic module of 3.3 m was used to divide the grid of the homestead; the widths of the main road and the alley were considered as 6.6 m and 3.3 m, respectively. The length of the standard homestead (H0) in the north–south and east–west directions were 13.2 m and 9.9 m, respectively, and the area was 130.68 m^2. The given base, as shown in Figure 5, comprised an abandoned gated factory which was surrounded by walls. Since this land is not residential, the scope of this was excluded from the design tests. It should be emphasized that centrality is an important factor in traditional Chinese rural settlement. Therefore, in the analysis of Space Syntax, the results of tests which had a higher integration of the central area were better.

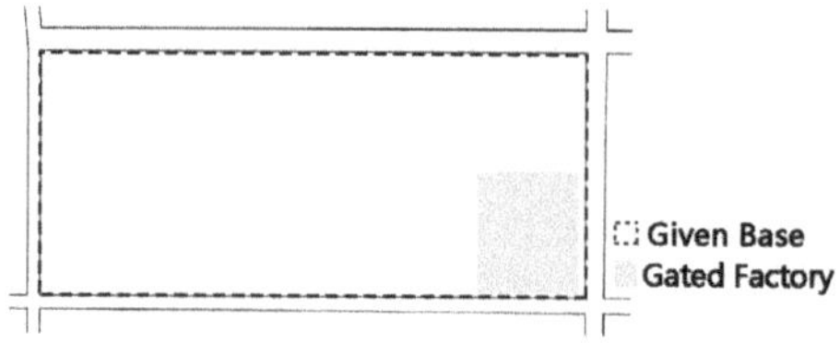

Figure 5. Given base.

5. Test 1 (Reappearance of General Mode of Planning)

Test 1 aimed to make the prevailing rural housing planning methods in China reappear. This process only involved the housing-level-based design rules, mainly by arranging and combining standard homesteads horizontally and vertically for layout.

5.1. Filling Rules

For a and b, starting from the left border, copy three (Figure 6a) or four (Figure 6b) of them horizontally to the right side and add a vertical road; repeat until the given base is filled. For c, starting from the upper border, copy two of them vertically to the lower side and add a horizontal road; repeat until the given base is filled (Figure 6c).

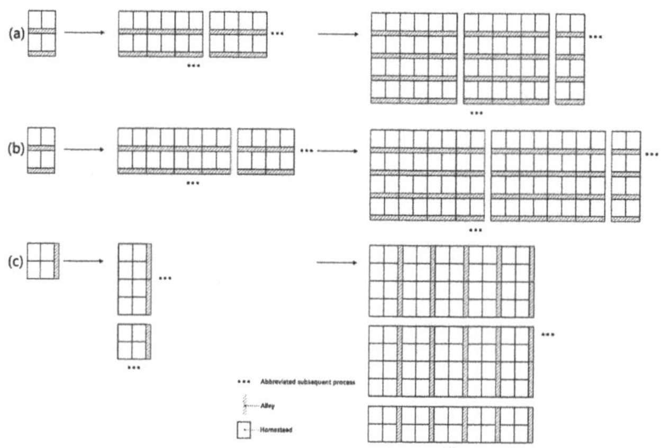

Figure 6. Filling rules. (**a**) The housing prototypes are arranged horizontally in groups of three. (**b**) The housing prototypes are arranged horizontally in groups of four. (**c**) The housing prototypes are arranged vertically in groups of four.

5.2. Implementation and Analysis

The filling rules were applied to the given base, generating three different plans to build the 3D models. Qualitatively, Figure 7a,b show the orderly form of row housing and homogeneous landscape, which displayed strong contrasts with the different styles and features of traditional villages. From a practical point of view, although this approach was conducive to mass production, the residential demands of each homestead could not be maintained as completely consistent, thereby losing the features of traditional villages. Quantitatively, while using Space Syntax to measure the integration value of the three results, the integration value showed an increase from the outer area to the central area, which indicated that the results were good with respect to integrity, centrality and enclosure (Figure 7c).

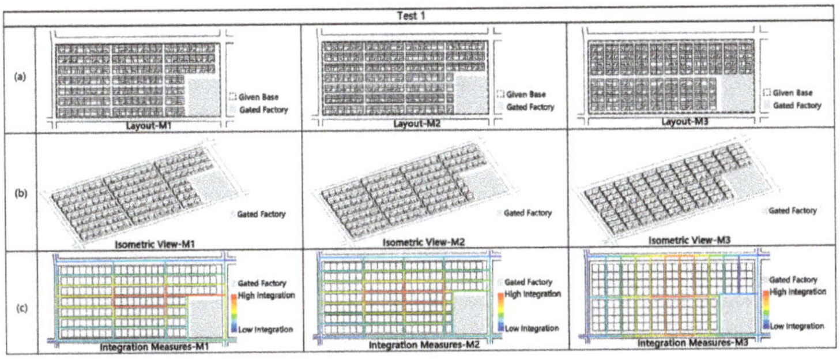

Figure 7. Results and analysis of Test 1. (**a**) Layout. (**b**) Isometric view. (**c**) The integration value of results measured.

6. Test 2 (Definition of SG-Based Rules)

More complicated rules were defined in Test 2 to address the problem of homogeneous design in Test 1. In order to improve the flexibility of the results, we not only defined the SG-based rules for housing level, but also defined the combination rules and filling rules for neighborhood and block level (Figure 8).

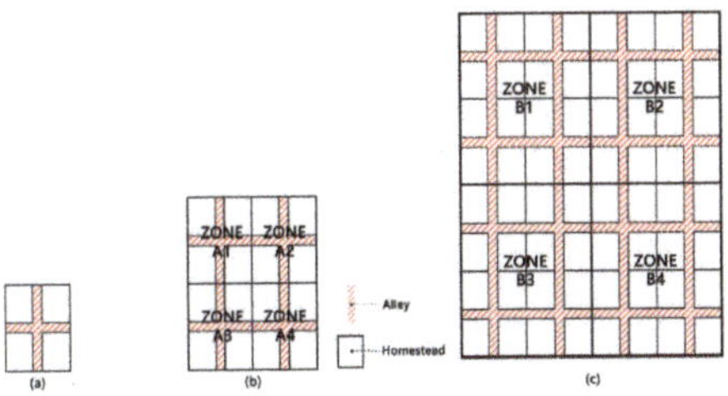

Figure 8. Prototype of neighborhood and block. (**a**) Housing prototype. (**b**) Neighborhood prototype. (**c**) Block prototype.

6.1. Housing Level: SG-Based Rules

In the housing prototype, the homesteads can be shifted horizontally or vertically by one module to form new shapes. Therefore, we used SG to define the rules to generate all the samples at the housing level. The main shape rules used are as follows (Figure 9):

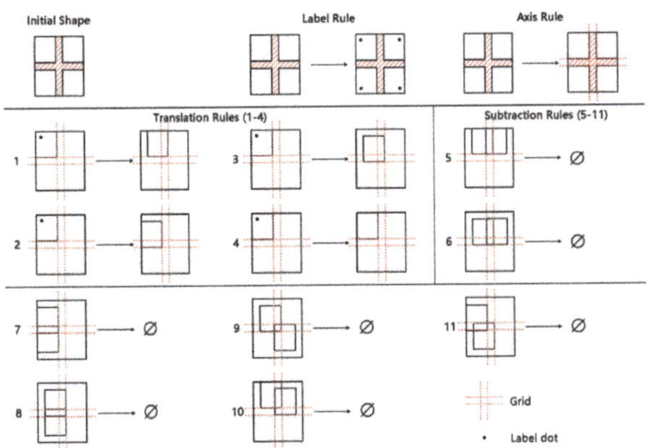

Figure 9. SG-based rules (Label rule, Axis rule, Transformation rules, Subtraction rules).

1. Label rule. Labeling homesteads of the initial shape indicated, on which the transformation rules could be applied.

2. Axis rule. This rule placed a grid over the initial shape and controlled the application of transformation rules, along with Euclidian transformations when translating the labelled (·) homesteads.

3. Transformation rules. The labels only allow for each of the homesteads to be acted on and can only be applied once to a selected homestead. Therefore, each homestead in the initial shape can only be manipulated once with the application of the four transformation rules.

4. Subtraction rules. As the purpose of the grammar was to facilitate samples of housing level development, the generation of overlapping or embedded shapes should be avoided. This is also supported by the application of subtraction rules which eliminate a complete permutation if overlapping or embedded shapes occur with the transformation of the smaller rectangles.

As permutations are progressively developed, they were categorized according to the number of applications of each of the four transformation rules. In the present study, the application of the shape rules and the subtraction rules generated 79 different samples of housing level—Set T (Figure 10).

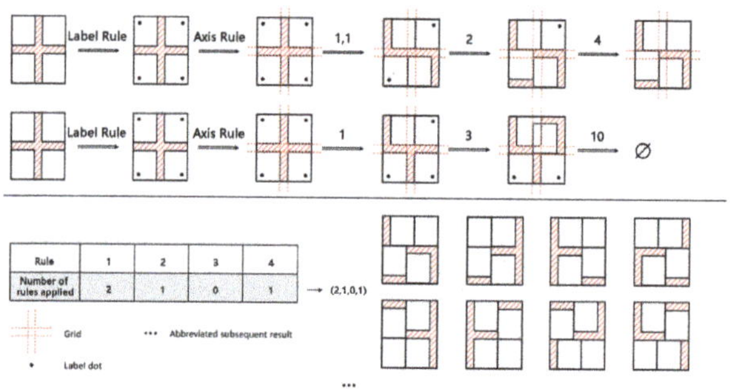

Figure 10. Implementation of rules.

6.2. Neighborhood and Block Level: Non-SG-Based Rules

1. Combination rules. The samples in Set T were randomly extracted and combined, like the neighborhood prototype, to form a new sample Set N. The samples in Set N were randomly extracted and combined, similarly to the block prototype, to form a new sample Set B.

2. Filling rules. The samples in Set B were collected and filled on the given base, and each sample was separated by a road. Moreover, during filling, 9 of the 64 homesteads were randomly transformed into open space.

6.3. Implementation and Analysis

We collected the samples from Set B and implemented the filling rules on the given base. Any extra homesteads beyond the boundary were removed to generate three different plans and build 3D models. Qualitatively, Figure 11a, b displays a repetitive and different form which has the traditional rural style and features. This shows good similarity and continuity in the results. Some shapes were identified from the traditional housing layout; however, there was only one size of homestead which did not meet the different residential demands. The results obtained were contingent, and reflected some errors that the gate could not be set up at the homestead (red-marked in Figure 11a). Quantitatively, the integration value of Test 2 was unevenly distributed within the given base, which means that the results did not show good centrality and enclosure, and the space was comparatively negative (Figure 11c). This is because some samples lacked the definition of restrictions or rules related to road connectivity in the combination process.

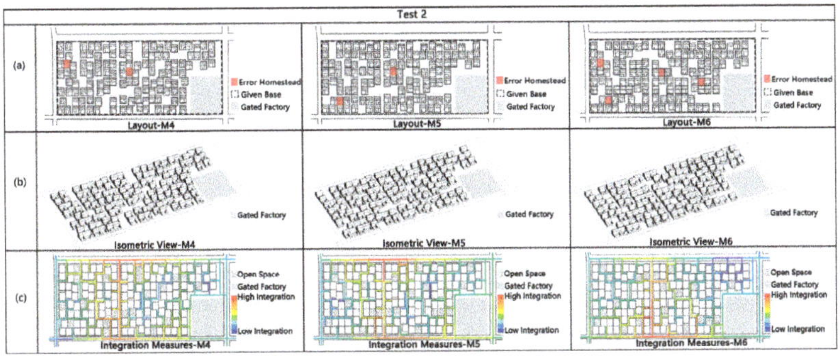

Figure 11. Results and analysis of Test 2. (**a**) Layout; (**b**) Isometric view; (**c**) The integration value of results maesured.

7. Test 3 (Optimization and Programming of Rules)

7.1. Predefinition

Based on the results of Test 1 and Test 2, the rules were optimized and coded using a computer program using Pyautocad. The samples in Set T and the housing and neighborhood prototypes were retained. Furthermore, to meet different residential demands, five sizes of homestead were predefined, including one standard type (H0), one small type (H1) and three large types (H2, H3, H4) (Figure 12a). Additionally, the block prototype was modified by adding north–south and east–west roads at the center (Figure 12b).

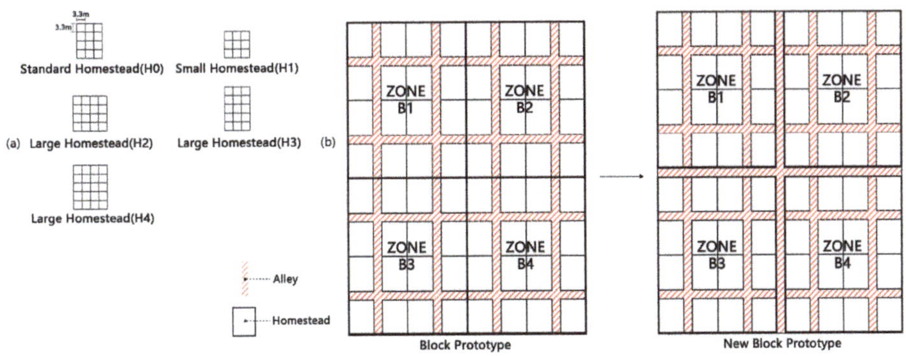

Figure 12. Adjusted prototypes of homestead and block. (**a**) Five sizes of homestead. (**b**) Modify block prototype.

7.2. Level-1 Rules (Housing Level)

1. Filtration rules. Limited by factors such as walking accessibility and gate position, Set T did not support the survey conclusions fully. Therefore, some samples were removed by setting new filtration rules, which are as follows: (1) as long as there is a homestead not close to an internal alley, filtering was necessary, (2) if the gate can only be opened on the north side of the homestead or the gable wall of the core housing, it must be filtered and (3) delete the sample of housing prototype. After filtering, 16 samples constituted the level-1 corpus (T0) (Figure 13a).

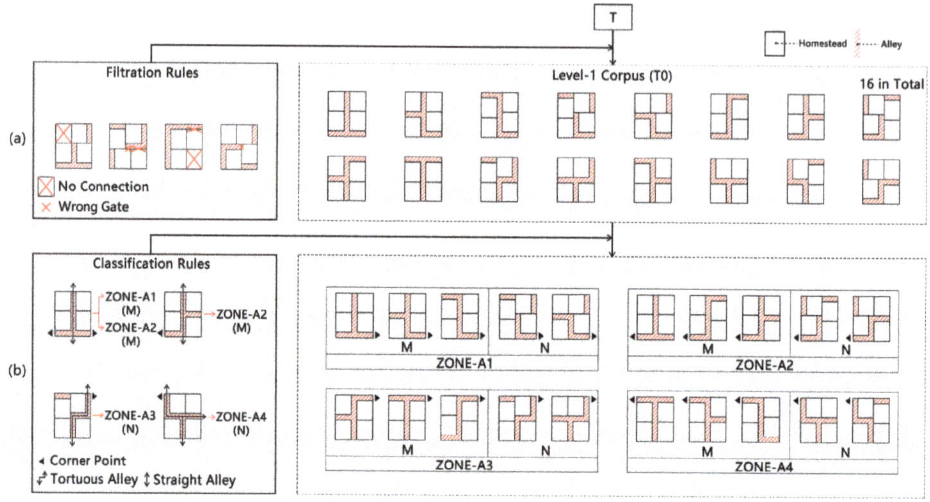

Figure 13. Level-1 rules. (**a**) Filtration rules. (**b**) Classification rules.

2. Classification rules. The level-1 corpus was divided into four quadrants for combination (ZONE-A1/A2/A3/A4), following which the four corner points could be accessed by the internal alleys. The samples that had access to the southeast corner point were placed in ZONE-A1, the ones with access to the southwest corner point were placed in ZONE-A2, while the ones with access to the northeast and northwest corner points were placed in ZONE-A3 and ZONE-A4, respectively. Since some samples' internal alleys could reach two corner points, they qualified to be placed in two quadrants equally. Also, to enrich the styles and features, the samples in each quadrant were divided into subsets M and N based on whether there were alleys running through the samples vertically (Figure 13b).

7.3. Level-2 Rules (Neighborhood Level)

1. Combination rules. First, in the neighborhood prototype, the level-1 samples of the four quadrants (ZONE-A1/A2/A3/A4) were extracted to fill. To avoid the monotonic problem caused by combinations, the following constraints were imposed during filling: (1) no adjacent ZONEs would have a mirror combination of the same samples and (2) no samples from M subsets of ZONE-A1 and ZONE-A3 can be adjacent to each other. Similarly, no samples from M subsets of ZONE-A2 and ZONE-A4 can be adjacent to each other. After imposing the above constraints, 148 combination shapes were obtained (Figure 14a).

2. Filtration rules. To improve accessibility, as long as one side of the neighborhood samples could not open the entrance to access the center, filtration became necessary. After filtering, 102 level-2 shapes were generated (Figure 14b).

3. Adjustment rules. Homesteads facing the main road often have a larger area, mainly used for selling daily necessities. There are also a small number of large homesteads in the inner areas not facing the road. Therefore, we used three larger homesteads (H2, H3 and H4), to accommodate the above possibilities. At this level, it is necessary to discuss the two cases of facing road and non-facing road. In advance, we should consider the following conditions: (1) a homestead at a corner point which did not face the main road should be selected as a labelled homestead; (2) the level-2 shapes with three or four sides facing the main road are not discussed. Furthermore, six types of adjustment rules were employed to generate the level-2 corpus, as described below.

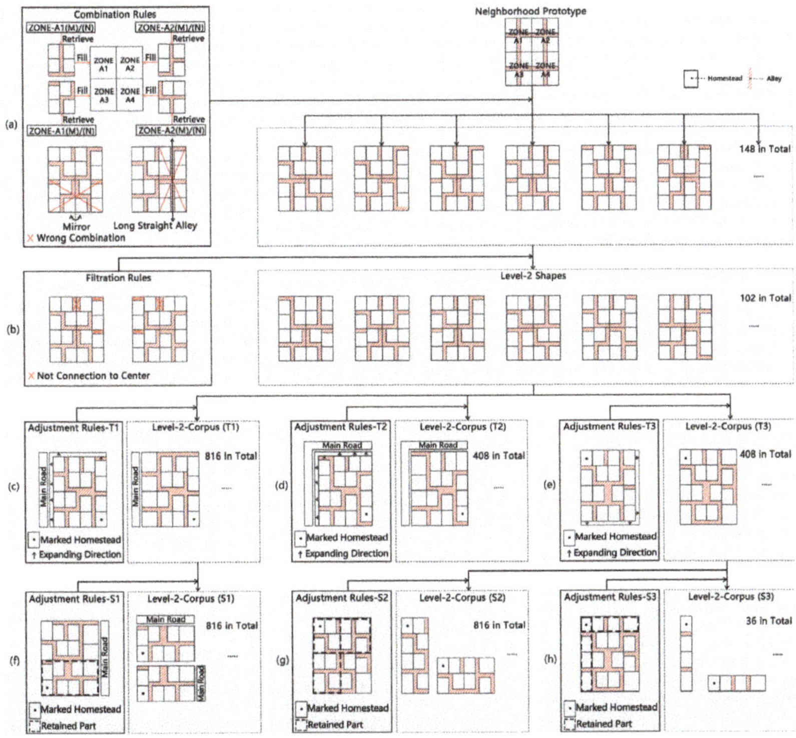

Figure 14. Level-2 rules. (**a**) Combination rules. (**b**) Filtration rules. (**c**) Adjustment rules—T1. (**d**) Adjustment rules—T2. (**e**) Adjustment rules—T3. (**f**) Adjustment rules—S1. (**g**) Adjustment rules—S2. (**h**) Adjustment rules—S3.

- Adjustment rules—T1. When only one side faced the main road, five standard homesteads turned into large homesteads, as shown in Figure 14c;
- Adjustment rules—T2. When two sides faced the main roads, seven homesteads became large homesteads, as shown in Figure 14d;
- Adjustment rules—T3. If no side faced the main road, three homesteads became large homesteads, as shown in Figure 14e;
- Adjustment rules—S1. Based on T1, the samples should be cut in half and the parts with the labeled homestead should be retained, as shown in Figure 14f;
- Adjustment rules—S2. Based on T3, the samples should be cut in half and the parts with the labeled homestead should be retained, as shown in Figure 14g;
- Adjustment rules—S3. Based on T3, the samples should be cut in quarters and the parts with the labeled homestead should be retained, as shown in Figure 14h.

Among these, the adjustment Rules S1, S2 and S3 are able to generate smaller samples of the neighborhood for filling the narrow or patchy space of a given base.

7.4. Level-3 Rules (Block Level)

The level-3 rules are the control rules for filling a given base to generate a block, so no level-3 samples were generated. We filled the block prototype facing the main road with the corpus of T1, T2 and T3 to verify how the samples of the level-2 corpus recognized the main road. Then, the following rules of reduction and adjustment control were set to recognize the number and area of homesteads in the block, and form of the block's open space.

1. Filling rules. When one side is facing a main road, the corpus of T1 and S1 can be used. When two sides are facing the main roads, the corpus of T2 and S1 can be used. When no side is facing the main roads, the corpus of T0, T3, S2 and S3 can be used. To improve the accessibility of traffic inside and outside the block, roads should also be reserved between the level-2 samples (Figure 15a).

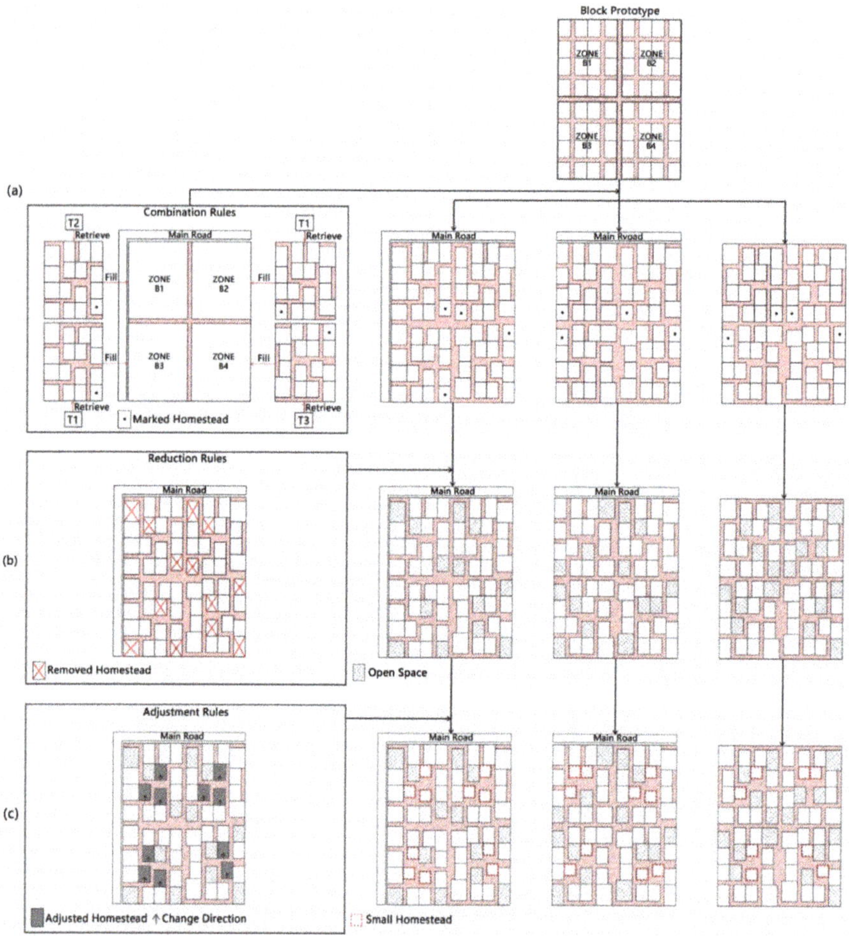

Figure 15. Level-3 rules. (**a**) Filling rules. (**b**) Reduction rules. (**c**) Adjustment rules.

2. Reduction rules. Open space is necessary for a block, so some homesteads must be removed. This includes the following two cases: (1) homesteads that are facing the main road could be block prototype entrances, and (2) homesteads that are not facing the main road could be open spaces for landscaping. Therefore, reduction rules are an effective method for dynamically adjusting the number of homesteads according to specific situations (Figure 15b).

3. Adjustment rules. To further increase the type of homesteads, standard homesteads (H0) at the center of a block can be transformed into small homesteads (H1) (Figure 15c).

7.5. Implementation and Analysis

By analyzing Test 1 and Test 2, more reasonable rules of the three levels can be redefined in Test 3 and coded using a computer program to generate the planning solutions.

These rules and the resulting corpus are more flexible and can adapt to more complex situations of construction land. For a given base, samples are collected from T0, T1, T2, T3, S1, S2 and S3 for filling according to these rules, and three different results (2D and 3D) were generated randomly (Figure 16a,b). We used Pyautocad to automate the design in AutoCAD. The specific operating principles of the program designed in Test 1-3 are as follows:

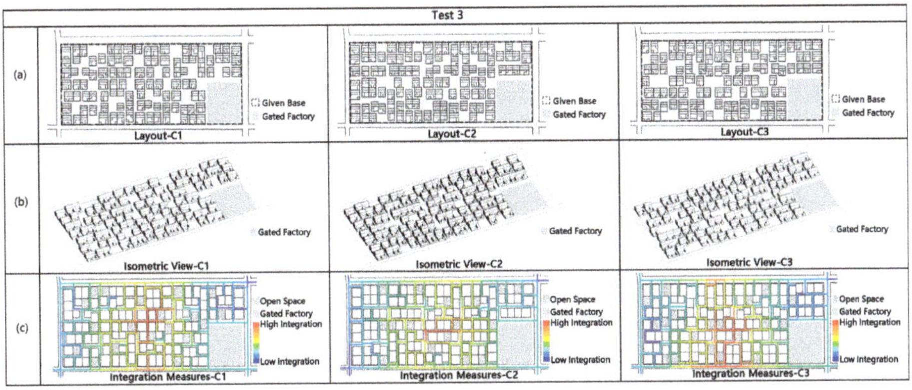

Figure 16. Results and analysis of Test 3. (**a**) Layout. (**b**) Isometric view. (**c**) The integration value of results measured.

1. Sample representation. As the module system was predefined in advance, a matrix of numbers can be used to represent the positional relationships between the homesteads, alleys and roads. For example, the level-1 samples were represented by a 9 × 7 matrix, where 1 represents the homestead and 0 represents the alley.

2. Sample filtration. In terms of filtration rules of level-2 rules, the sample matrices were accessed one by one using Depth-First Search (DFS), complying to the rules which allowed us to identify the unreasonable samples, such as one side of the neighborhood samples cannot open the entrance to access the center.

3. Sample combination and classification. During the combination process, the low-level samples were classified first via DFS, and the prerequisites for the high-level combined samples were analyzed using algorithms. The low-level samples were then combined with the high-level ones following matrix combinations.

4. Homestead identification and filling. When the rules were applied to a given base, the level-2 corpus should be used first, identified and filled according to the corresponding layers (base boundary and main road). Then, the level-1 corpus can be called to fill the remaining small space, in order to complete adaptive filling at the given base. Moreover, in response to the design goals, some parameters were set to control the results. For example, to render the main road accessible, the filling probability of the S1 and S3 samples can be increased to identify the boundary of the base and exclusive layer; the filling probability of the T3 and S2 samples can be increased.

Qualitatively, the results of Test 3 were similar to those of Test 2, showing the appearance of traditional rural settlements. Quantitatively, the results of Test 3 were similar to those of Test 1. The integration value of these results was higher for the central area, which meant that the central area had high accessibility (Figure 16c). In general, the results of Test 3, based on automatic design, better represented the characteristics of traditional rural self-organization and this is more reasonable in road planning.

8. Discussion

The three tests, T1, T2 and T3, applied to the given base, with a small area and relatively regular shape, were found to be suitable for the rapid qualitative and quantitative analysis of the results of reappearing the characteristics of traditional rural settlements. However, due to the regular shape and small size of the given base, it could not be proved that a feasible layout could be generated when dealing with a larger and more irregular site. In addition, due to a small number of samples collected at all levels, it was impossible to show the mutual influence of a large number of samples at all levels during filling. As a result, to further verify the adaptability of the application of this design approach, we selected a new base (Figure 17a) for a test which had more complex site conditions. The corpus and rules of Test 3 were used. The main road and site boundary in AutoCAD were automatically identified by the computer program (Figure 17b), and the layout of the homestead was also predefined and randomly generated (Figure 17c). Both qualitatively and quantitatively, the results of the test successfully retained the traditional Chinese rural style and characteristics (Figure 17d, e). Moreover, the random filling and mutual influence of a large number of samples at all levels created a flexible and diverse road network at the block level (Figure 17b). Therefore, the test proved that the corpus and rules can reasonably and accurately generate the housing layout in accordance with the traditional rural characteristics, even on a large-sized irregular site. However, due to the size limitation of the samples, some small areas could not be filled (Figure 17b, c). In addition, since the samples have regular rectangular boundaries, they may not be well adapted to sites with various actual complex conditions, such as curved or oblique boundaries. In future work, we intend to improve the three-level rules, enrich the corpus and filling rules and make them adaptable to more complex topographic conditions, such as altitude difference and irregular site boundary. Moreover, since the main roads and alleys are only considered in this work, we plan to improve the communal-facilities-based rules corpus at the settlement level and town level, with a view to developing the generative approach.

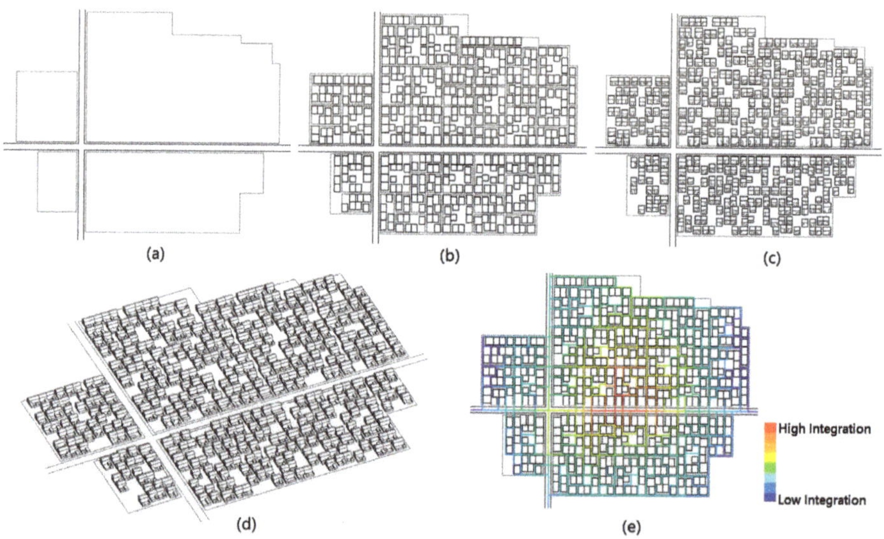

Figure 17. The results and analysis of one new test. (**a**) A new base. (**b**) Filling automatically. (**c**) Layout. (**d**) Isometric view. (**e**) The integration value of result measured.

9. Conclusions

This study explores a rule-based design approach for rural mass housing, mainly based on the results of three tests. The optimized rules and condition restrictions were coded, and a relatively reasonable result was obtained via the automatic operation of the program. Through a comprehensive analysis of the tests, the following conclusions can be drawn. First, this rule-based generative design approach can help architects solve the design problem of the homogenization of rural mass housing, and the design results have "harmony in diversity" characteristics. The method keeps architects' role in the housing design process. It acts as a research booster, helping architects systematically explore the logic and process of generating research objects. Secondly, compared with traditional design methods, it can achieve the "emergence" and "unpredictability" of design results through rule-based reasoning, which can better meet the needs of multiple parties, especially associating architects and users. It can not only help architects use computer-aided tools to select and optimize the generated results and better control the geometric size, entrance direction and other parameters to realize the traceable and targeted research process, but also allow users to adjust the layout of their houses and their built-in building information according to their needs, depending on the family's life cycle. Thirdly, as a visual design tool based on shapes, SG has the advantages of processing multiple complex rules simultaneously, directly applying rules to functional geometry and better perceiving and expressing complex shapes. An effective relationship between abstract "spatial demands" (traditional architectural features, user design demands) and specific design rules can be established. Architects can apply this approach to control the design generation process, optimizing the design quality and improving the design efficiency.

10. Patents

The following Chinese invention patent was produced: "A Generative Design Method for Mass Housing Layout in Rural Areas of the North China Plain", with the authorization number "ZL202010327152.8".

Author Contributions: Idea, Methodology and Survey Data: J.W., W.F., B.Z. and Z.Z.; Model Analysis: J.W., Y.Y. and W.F.; Rules Definition and Verification: J.W., W.F. and B.Z., Computer Programming: Z.Z. All images are by the authors. All authors have read and agreed to the published version of the manuscript.

Funding: This work was supported by the Shandong Natural Science Foundation Project (ZR2021MF099, ZR2022MF334); the Shandong Province Teaching Reform Research Project (M2021130, M2022245, Z2022202); the Shandong Province High Quality Professional Degree Teaching Case Library Construction Project (SDYAL2022155).

Data Availability Statement: All data used in this research are available on open data sources.

Acknowledgments: We would like to give special thanks to K. Liu, L. Qiao, S. Sang and Q. Q. Qin for their support.

Conflicts of Interest: The authors declare no conflict of interest.

References

1. Wang, J.; Zhao, J.; Wu, T.; Li, J. A co-evolution model of planning space and self-built space for compact settlements in rural China. *Nexus Netw. J.* **2017**, *19*, 473–501. [CrossRef]
2. Oxman, R. Theory and design in the first digital age. *Des. Stud.* **2006**, *27*, 229–265. [CrossRef]
3. Singh, V.; Gu, N. Towards an integrated generative design framework. *Des. Stud.* **2012**, *33*, 185–207. [CrossRef]
4. Rahimian, M.; Beirão, J.N.; Duarte, J.P. A grammar-based generative urban design tool considering topographic constraints-the case for American urban planning. In Proceedings of the 37th eCAADe and 23rd SIGraDi Conference, Porto, Portugal, 11–13 September 2019.
5. Stiny, G.; Gips, J. Shape grammars and the generative specification of painting and sculpture. In Proceedings of the IFIP Congress, Ljubljana, Slovenia, 23–28 August 1971.
6. Stiny, G.; Mitchel, W.J. The Palladian grammar. *Environ. Plan. B Plan. Des.* **1978**, *5*, 5–18. [CrossRef]
7. Chiou, S.C.; Krishnamurti, R. The grammar of Taiwanese traditional vernacular dwellings. *Environ. Plan. B Plan. Des.* **1995**, *22*, 689–720. [CrossRef]
8. Duarte, J.P. Towards the mass customization of housing: The grammar of Siza's houses at Malagueira. *Environ. Plan. B Plan. Des.* **2005**, *32*, 347–380. [CrossRef]
9. Beirão, J.N.; Duarte, J.P.; Stouffs, R. Creating specific grammars with generic grammars: Towards flexible urban design. *Nexus Netw. J.* **2010**, *13*, 73–111.
10. Wang, J.; Zhao, B.; Fan, W.; Yang, Y.; Zhao, J. A combined shape grammar and housing-space demand approach: Customized mass housing design in rural areas of the North China Plain. *Nexus Netw. J.* **2022**, *24*, 5–23. [CrossRef]

11. Duarte, J.P. Customizing mass housing: Toward a formalized approach. In *Mass Customization and Design Democratization*, 1st ed.; Kolarevic, B., Duarte, J.P., Eds.; Routledge: London, UK, 2018; pp. 129–142.
12. Wang, J.; Zhang, S.; Fan, W. Application of shape grammar to vernacular houses: A brief case study of Unconventional Villages in the contemporary context. *J. Asian Archit. Build. Eng.* **2023**. [CrossRef]
13. Celani, G. Grammar form and analysis methodology and synthesis in architecture. *Connect. Commun. Cult.* **2006**, *5*, 180–197.
14. Erem, Ö.; Abbasoğlu, E.M.S. Adapted design generation for Turkish vernacular housing grammar. *Environ. Plan. B Plan. Des.* **2016**, *43*, 893–919. [CrossRef]
15. Herbert, T.; Sanders, I.; Mills, G. African shape grammar: A language of linear Ndebele homesteads. *Environ. Plan. B Plan. Des.* **1994**, *21*, 453–476. [CrossRef]
16. Dias, M.A. Informal settlements: A shape grammar approach. *J. Civil Eng. and Archit.* **2014**, *8*, 1389–1395.
17. Jiang, B.; Claramunt, C. Integration of space syntax into GIS: New perspectives for urban morphology. *Trans. GIS.* **2002**, *6*, 295–309. [CrossRef]

buildings

MDPI

Article

Towards an Evaluation System of Disabled Individuals' Friendly Communities from the Perspective of Inclusive Development—A Case Study in Jinan

Yaqi Wen, Yuyao Li, Yang Yang * and Jiang Wang

College of Architecture and Urban Planning, Shandong Jianzhu University, Jinan 250101, China
* Correspondence: yangyang21@sdjzu.edu.cn

Abstract: The quality of travel for residents has improved in China's urban renewal development, but there are still barriers to the activities of disadvantaged groups like people with disabilities in the city, including challenges with accessible travel and a lack of accessible design. All urban people should have access to services and be guaranteed equal rights, according to the inclusive development theory. This study examines the travel requirements and behavioural patterns of the barrier-free groups to safeguard the right of underprivileged groups to participate equally in the urban space. It also determines the impact of various types of urban land use and service facilities on the distribution of barrier-free facilities. An evaluation system based on the needs of the disabled group is established to evaluate the degree of barrier-free facilities in the surveyed areas, summarise the problems shown by the evaluation scores, and propose improvement strategies for the shortcomings of accessibility to residential areas and the controversy through web data crawling and a geographically weighted analysis. From the standpoint of inclusive development, the goal is to lessen the travel issues faced by underprivileged groups and to improve their capacity to take advantage of opportunities and have access to resources.

Keywords: inclusive development; disability friendly; accessible facilities evaluation; Jinan

1. Introduction

1.1. Backgrounds

At the end of 2022, the urbanisation rate of China's resident population was 65.22%, which has stepped into the middle and late stages of the rapid development of urbanization. The development of urbanization has entered an important period of urban renewal, with the development path switching from "quantity" to "quality", and from large-scale incremental construction to stock-quality renovation focusing on improving urban quality [1]. In the context of urban renewal, the 2030 Agenda for Sustainable Development puts forward goals for building inclusive and sustainable cities, including the following goal: by 2030, to strengthen inclusive and sustainable cities and capacities for participatory, integrated, and sustainable human settlements planning and management in all countries and to provide universal access to safe, inclusive, accessible, and green public spaces for all, in particular, for women, children, older persons, and persons with disabilities [2]. Where the development of urban accessibility will immediately affect the travelling conditions of the elderly and persons with disabilities, etc., the construction of an accessible travelling environment is, therefore, an important part of inclusiveness. The 14th Five-Year Plan emphasises strengthening service facilities for persons with disabilities and comprehensive service capacity building, and improving the construction of barrier-free environments [3]. The State gives considerable attention to vulnerable groups, and the necessity to improve the barrier-free environment opens the way to the building of inclusive cities.

People with disabilities are a typical vulnerable group in society, and their needs must be seriously considered in the process of raising the quality of urban development.

More than 85 million Chinese citizens are disabled, and there are more than 64 million senior individuals who are disabled or semi-impaired. The design of accessible travel has emerged as a key issue in urban redevelopment since there is a sizable but unmet demand for it and significant challenges in the areas of basic life safety and social participation.

China began to build an accessible environment relatively late, with the concept of "accessibility" appearing for the first time in 1985 in a seminar on "Persons with Disabilities and the Social Environment". Some scholars have concluded that although the barrier-free design in China is being gradually improved, at the same time, there are still a lot of places that need to be improved: from the perspective of society, the designers do not have a sufficient understanding of accessible facilities, and all parties do not attach enough importance to the research and construction of the design of barrier-free access; from the perspective of management, there are often cases where barrier-free access is piled up with sundry things or damaged by road construction [4]. From a theoretical point of view, the overall research is still insufficiently emphasised, and there is a lack of better and more systematic design methods [5]. Overall, China's construction of an accessible travelling environment is insufficient because of the lack of attention to the perspectives of persons with disabilities and the psychology of persons with disabilities, as well as the lack of quantitative statistics and quantitative analyses of the degree of uniformity of the construction of accessibility facilities.

A nation's ability to advance economically and socially depends heavily on its infrastructure. However, it is crucial to evaluate the condition of the infrastructure, determine the regions that require the greatest care, and allocate the government's constrained budget to those areas [6]. More specifically, it would be effective and crucial to evaluate the present state of accessibility in a city to identify the areas that require updating and more relevance to the actual needs of people with disabilities.

The inclusive design of urban space is inextricably linked to the social component, which emphasises the city's openness, acceptance, recognition, and respect for all groups and classes of residents [7]. Current divisions should be improved to reduce the exclusion and segregation of disadvantaged groups in urban spaces, to guarantee equal power for every resident, and to promote equity and justice [8,9]. This paper analyses the travel needs and behavioural patterns of the accessible groups based on the theory of inclusive perspectives, focusing on the needs and problems of disadvantaged public groups in the context of urban regeneration and urban construction. The goals of this paper are to improve the quality of urbanisation, guarantee the right of disabled groups to use urban resources equally, and reduce the travel problems of disabled people. It proposes improvement strategies for the management of barrier-free facilities in combination with the needs of disabled groups, measuring the degree of perfection of the barrier-free environment through the establishment of an assessment system for the degree of perfection of barrier-free road facilities, to draw attention to the underprivileged groups and increase their opportunities for participation in public life.

1.2. Research Framework

Based on previous research, the urban accessibility update section is deepened, and quantitative analyses of the degree of urban accessibility improvement are conducted, an evaluation system is established, and the results of the research on the needs of persons with disabilities are made more intuitive, so as to assist in the analysis of the subsequent data and provide the basis for urban renewal. It then proposes urban renewal strategies and recommendations to make urban renewal services better and more inclusive. Figure 1 shows the flowchart of the paper's research.

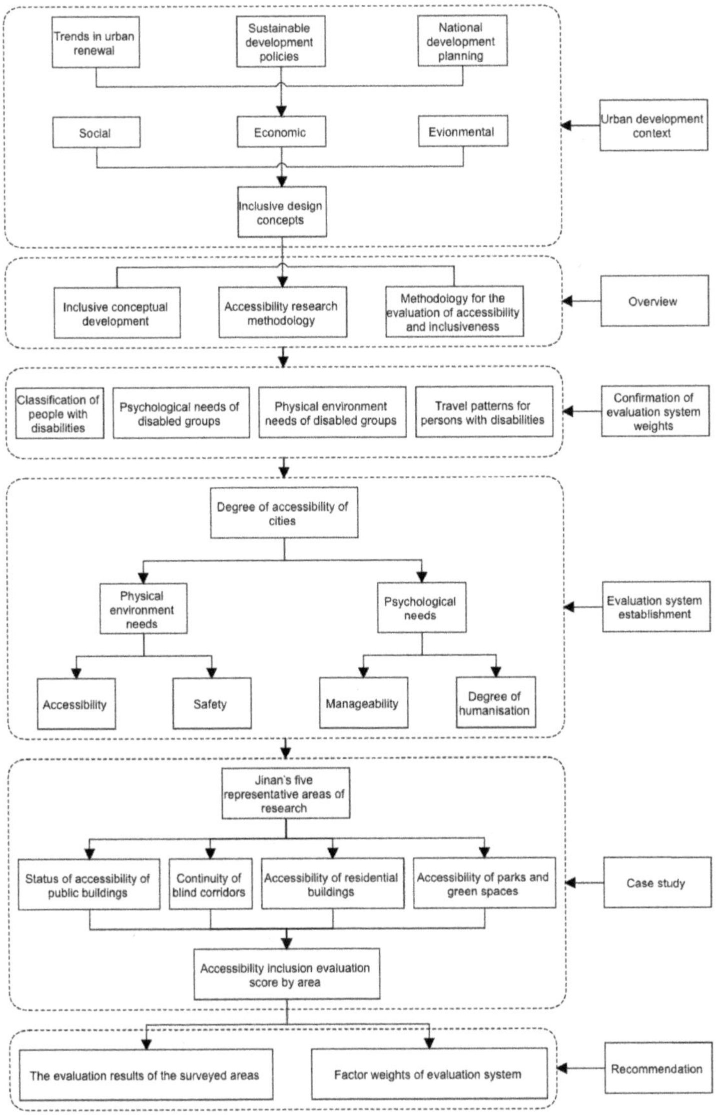

Figure 1. Research flow chart (source: the authors).

2. Overview

2.1. The Definition and Evolution of Inclusive Design Concepts

Inclusive development involves various aspects. From a social perspective, UN-Habitat first formally proposed the "inclusive city" initiative in the *World Cities Report* in 2000, emphasising the equal rights, resource sharing, and spatial balance of different urban subjects in the process of urban development, and promoting the equal participation of urban citizens in the development process under the guidance of the concept of inclusiveness; at the 2016 United Nations Conference on Housing and Sustainable Urban Development, inclusiveness was incorporated into the New Urban Agenda as a core concept, with "Cities for All, Equal Use and Enjoyment of Cities and Human Settlements by All" as a common vision to promote prosperity and improve the quality of life for all, by promoting social cohesion, inclusiveness, and security in peaceful and pluralistic societies in which the needs of all inhabitants are met and the special needs of those in vulnerable situations are recognised [10]. The obstacles to travelling by vulnerable groups, such as the disabled and the elderly, should be eliminated and their right to participate in normal life should be guaranteed.

The notion of inclusive growth was initially introduced by the Asian Development Bank (ADB) in 2007 from an economic standpoint. Its objective was to enhance the accessibility of impoverished individuals to markets and essential productive assets, as well as to ensure their meaningful involvement in decision-making regarding resource management [11]. Within the framework of China's urban development agenda, inclusive growth entails the equitable distribution of economic development benefits throughout the urban development process, fostering coordinated and balanced development, and ultimately achieving shared prosperity [12].

In the spatial environment perspective, inclusive thinking first appeared in 1984 when Richard Hatch introduced the idea of inclusive design, defining it as "the ability of the public to participate in and control the environment". Later in 1994, Roger Coleman proposed inclusive design as an offshoot of disability design to create a friendlier future for all in terms of environmental design and assessment [13].

2.2. *An Overview of Accessible Design from the Perspective of Inclusive Development*

2.2.1. Introduction to Research Methods in Accessibility Studies

Previous scholarly investigations on accessibility design have employed various research methods such as questionnaires, simulations of disabled individuals' travel experiences, analysis of population characteristics, user research through zonal random sampling surveys, and others. These studies have focused on examining the current issues related to accessibility, the satisfaction levels of disabled populations, the challenges posed by obstacles, and the fairness of participation in social activities. Additionally, these studies have proposed specific strategies to address these issues. However, it is important to acknowledge that these investigations have certain limitations. For instance, they tend to overlook other types of disabilities when studying a specific group of disabled individuals. Moreover, these studies heavily rely on subjective human judgement and a qualitative analysis, lacking a quantitative analysis and the establishment of a comprehensive evaluation system. Specific literature arrangements are summarized in Table 1.

Table 1. Accessibility research methods review form.

Target	Methodology	Contents	Strategies
Utilisation rate of accessible facilities [14]	In-depth interviews and fieldwork	Socio-economic characteristics of persons with disabilities, travelling activities, living space, facility utilisation influencing factors	Universal design, optimising accessibility, frequency of use
Continuity and safe accessibility for accessible travelling [15]	Questionnaires, simulated trips for people with disabilities	Sample survey on travelling by people with disabilities, status and problems	Safeguard and regulatory system, promoting information accessibility
Status of facilities in the quarter-hour community service circle [16]	Questionnaire, factor analysis of population characteristics	Characteristics of the physically disabled population, characteristics of place needs, correlation between environmental needs and satisfaction	Addressing systemic issues, improve the education level of people with physical disabilities.

Table 1. *Cont.*

Target	Methodology	Contents	Strategies
Ordinary citizens, sanitation workers, the visually impaired [17]	Random sampling by district	Restrictions on the use of blind alleys, status of low travelling rate of visually impaired persons	Strengthening publicity, strengthening construction and maintenance, provide rehabilitation training for the visually impaired
Travelling life patterns of people with disabilities [18]	User research, environmental studies	Participation in social equity, assistive technology for travelling	Incorporate assistive technology for travelling

2.2.2. Introduction to Inclusive Evaluation

The evaluation of urban inclusiveness is a further deepening of the evaluation of accessibility. The design of accessible facilities necessitates the involvement and assessment of disability groups to ensure that their intentions and needs are comprehended and fulfilled, thereby fostering a more participatory and inclusive approach. Previous suggestions have emphasised the importance of recognising the diversity within the disabled community, promoting understanding and respect for individuals with disabilities, facilitating their active participation and cocreation in the design process, and conducting participatory observations to assess the actual usability of facilities by people with disabilities as part of the overall quality control. This research paper aims to contribute to the development of an evaluation system by investigating the specific usage requirements of individuals with disabilities. Relevant literature are summarized in Table 2 below.

Table 2. Inclusive accessibility design review form.

Concept Proposal	Participation of the Disabled	Methodology
Promote inclusive development and strengthening systematization [19].	Processes of policymaking	An analytical framework to study design and processes
Create community participation and boundary objects [20].	Participant observation Processes participation and cocreation	Two participatory agenda-setting projects
Recognize the diversity of disabilities. Expand the participation of marginalized groups in tourism decision-making [21].	Tourism producers, tourism consumers	Applying the conceptual framework for inclusive tourism development
Explore innovative barrier-free design methods [5].	Interaction with the design team	Conceptual and developmental overview Generalisation of methods and theoretical models
Inside the barrier-free community participatory planning [22]	Participation in community public hearings	Pattern language

3. Evaluation System

3.1. Establishment of the Evaluation

After consulting the pertinent policy manuals and papers in a previous article, it is necessary to research and compile the travel requirements and behavioural patterns of the barrier-free groups to obtain the corresponding and more accurate impact indicators and weighting values. This will help with the renewal and construction of barrier-free facilities as well as the evaluation of the impact of various types of urban service facilities.

3.2. Needs of the Disabled Population

The direct aim of accessibility is to solve the mobility problems of disabled people, and through literature references, the needs of disabled people were collated from different perspectives in order to serve as a basis for the establishment of evaluation factors.

As users of barrier-free facilities and as special people in society, persons with disabilities need a change in perspective, the full consideration of their needs, the satisfaction of their needs as much as possible, and assistance in having their demands "seen and heard". Persons with disabilities have the same basic and generic needs as all human beings and the same needs to be met. Thus, in conjunction with Maslow's theory of needs, the analysis focuses on the integration of the five areas of physiology, safety, socialisation, respect, and self-actualisation [23]. The introduction and ranking of these five aspects are shown in Figure 2.

Figure 2. Maslow's hierarchy of needs.

Firstly, the most basic physiological needs for people with disabilities are equivalent to their reliance on accessible facilities in social and urban construction, which requires designers to accurately plan the application and layout of accessible facilities in public facilities. Secondly, the same low-level safety needs for people with disabilities are the need for the safe use of accessible facilities and a safe arrival at destinations, and to ensure their safety, it is necessary to focus on the need for a regular maintenance, a reasonable design, the ease of use, and the effective use of public facilities. Social needs and respect needs are relatively high-level needs which belong to psychological aspects. The highest level of self-fulfilment needs refers to the process by which the individual's various talents and potentials can be brought into full play in a suitable social environment and the individual's ideals and ambitions can be realised, and in the case of a person with disabilities whose physical disabilities have already become a relative limitation, the limitations on their inner pursuits should be reduced to a minimum.

Based on this theory, and taking into account the realities of the situation, evaluation factors were determined on the basis of the needs of the different categories of the disabled groups themselves, the needs of the physical environment of the disabled, the psychological needs of the disabled, and their travelling patterns.

3.2.1. Categorization of Individuals with Disabilities

Through cooperation with the Jinan Disabled Persons' Federation, theoretical training, practical research, and classification of disabled groups were completed and the types of disabilities were categorised into four types: physical (A1), visual (A2), hearing (A3), and cognitive (A4) [24].

Persons with disabilities have different levels of disability and have different requirements for environmental accessibility. Among them, persons with physical disabilities who use mobility aids and wheelchairs and persons with visual disabilities have higher accessibility requirements: Persons with physical disabilities may have some difficulty in walking, climbing steps or slopes, and standing for long periods; persons with visual disabilities who have functional impairments in their vision or field of view mostly rely on their senses of touch and hearing to perceive their surroundings, or they use crutches or service animals to assist them in getting around. People with cognitive disabilities due to slow intellectual development have higher accessibility requirements, may have difficulty in using various service facilities, and have a limited range of activities, mainly in the residential area. Therefore, the main needs of this group of people are accessible facilities with strong environmental features and simple and clear flow lines in residential areas. On the other hand, persons with hearing disabilities and persons with upper-limb disabilities, who have difficulties in verbal communication, do not have walking difficulties, and have a high degree of self-care, have less issues with the accessibility of the built environment in terms of mobility, have lower accessibility requirements for the environment, and mainly require sufficient light and a quiet acoustic environment for hearing aids.

Through the categorisation study of different disability groups, the factors affecting the mobility of people with disabilities were compiled in descending order of influence: the continuity of blind corridors, the presence of ramp connections at height differences, the simplicity and ease of use of barrier-free facilities, the flow lines within the living area, and the light and acoustic environments.

3.2.2. Physical Environmental Requirements of Inclusive Communities

The physical demand primarily pertains to the need for accessible facilities and buildings, necessitating the implementation of appropriate design and management standards to address the physiological and safety requirements of individuals with disabilities. This encompasses various aspects, such as the inclusion of accessible entrances in residential buildings, the provision of flyovers and zebra crossings on pavements, and the construction of blind alleys within residential areas. Additionally, specific considerations must be made for certain facilities, such as designing wider accessible public toilets to accommodate wheelchair access and appropriately labelling parking spaces as barrier-free in car parks [25].

Travelling safety is the common mentality of all participants in urban transport, and it is the most basic requirement for all travelling activities [26]. Compared with ordinary people, disabled people are usually less capable of self-protection in traffic accidents than ordinary people due to their physiological insufficiency, so the probability and degree of injuries are higher than those of ordinary people.

The accessibility of information and communication also satisfies the physiological and safety requirements of people with disabilities. This is primarily reflected in the media facilities in public spaces that can communicate information to people with disabilities, enabling people with various types of disabilities to obtain information, primarily necessitating the consideration of both the visual and auditory senses [25]. Therefore, the weights of transport station services, public management and service facilities, and parks and green spaces should be moderately increased. Affected by the demand for community services and the shortcomings in accessibility services and awareness, the demand for accessible facilities in public utilities, commercial service facilities, and parks should also be moderately increased.

3.2.3. Psychological Requirements of Demographically Inclusive Cohorts

The user's control over the environment and the state of things is called controllability. Among them, psychological security is an important influencing factor [26]. For example, the continuity of the distribution of accessible facilities in transport venues and the availability of simple and easy-to-use accessible facilities affect the psychological needs of people with disabilities. Satisfying these psychological needs and providing a stable and humane experience for people with disabilities during travelling is conducive to helping people with disabilities overcome or even eliminate some of their psychological barriers, which in turn increases the likelihood of travelling in the future [26]. Among the previous needs by disability group, whether the barrier-free facilities are simple and easy to use, the flow line, light, and acoustic environment in a residential area can be used as controllable travelling factors, and the degree of universality of barrier-free facilities and the degree of appropriateness for different groups of disabled people can be used to evaluate whether the facilities are humanised or not.

3.2.4. Travel Behaviours Exhibited by Individuals with Disabilities

Accessibility refers to the ease of getting from one location to another and is often used to describe how human activities can maximise access using minimal activity and is an important basis when public facilities are built and planned [27]. Many factors in the city affect the accessibility of travelling for people with disabilities, and the higher the frequency of travelling, the higher the demand for this factor by people with disabilities, and the more important it is in the evaluation system. The research used data-collation field visits to the disabled and online questionnaires to summarise the travel patterns of the disabled; a total of 200 questionnaires were distributed, and the results showed that leisure trips and shopping trips accounted for the highest proportion of travel frequency, so when assigning weights affecting the distribution of accessible facilities, the proportion of parks, green spaces, and commercial service facilities should be increased as appropriate.

Based on the data shown in the aforementioned chart about the activity requirements of individuals with disabilities, it is evident that there is a greater occurrence of leisure and shopping excursions; however, the likelihood of engaging in other forms of travel is comparatively diminished. These results are summarized in Figures 3 and 4.

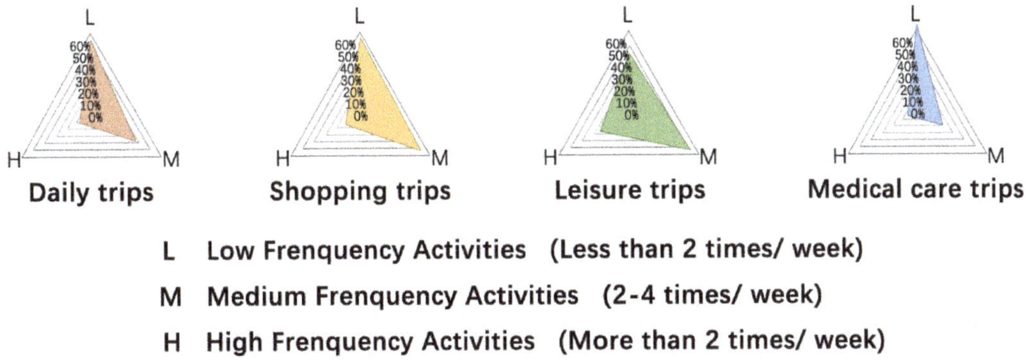

Figure 3. Radar map of frequency of travelling for different types of activities for persons with disabilities (source: the authors).

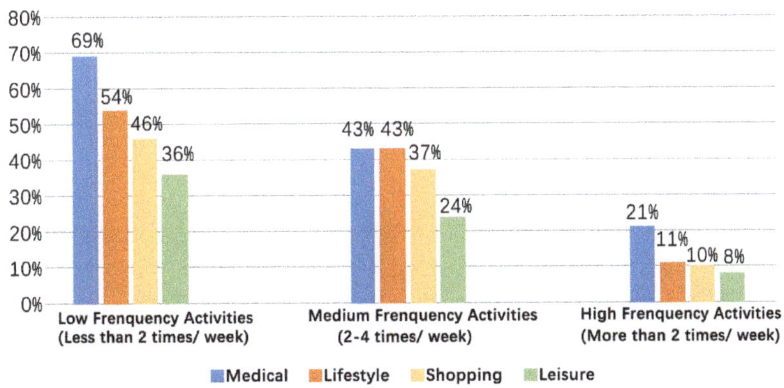

Figure 4. Histogram of travelling to different types of activities by persons with disabilities (source: the authors).

The chart illustrates the distribution of activity trips for disabled individuals across various categories. It reveals that medical and life service facilities exhibit the highest proportion in the likelihood of low-frequency trips. In terms of medium-frequency trips (2–4 days per week), both medical and life service facilities, as well as shopping facilities, demonstrate the highest proportion. As for high-frequency trips (more than 4 days per week), the proportion of medical, life, shopping, and leisure categories collectively decreases, although medical facilities remain predominant.

In conjunction with the previous section, public services and buildings for high-frequency travel were included in the accessibility factor for public buildings. Considering the constraints associated with travel distances for individuals with disabilities and the recommendations regarding the organisation of three-tier living areas, the public service facilities should consist of a blend of medical service facilities and living service facilities that are conveniently accessible to individuals with disabilities who travel by foot or wheelchair. These facilities should be accorded the highest priority compared with other service facilities.

3.2.5. Assessment of the Accessibility Scenario

In brief, the assessment of the level of adequacy of accessible facilities was conducted based on two dimensions: physical environment requirements and psychological needs. The former encompassed accessibility within buildings, road traffic safety, and the effectiveness of information exchange. The latter pertained to controllability and humanization. The specific factors for the evaluation are depicted in Figure 5.

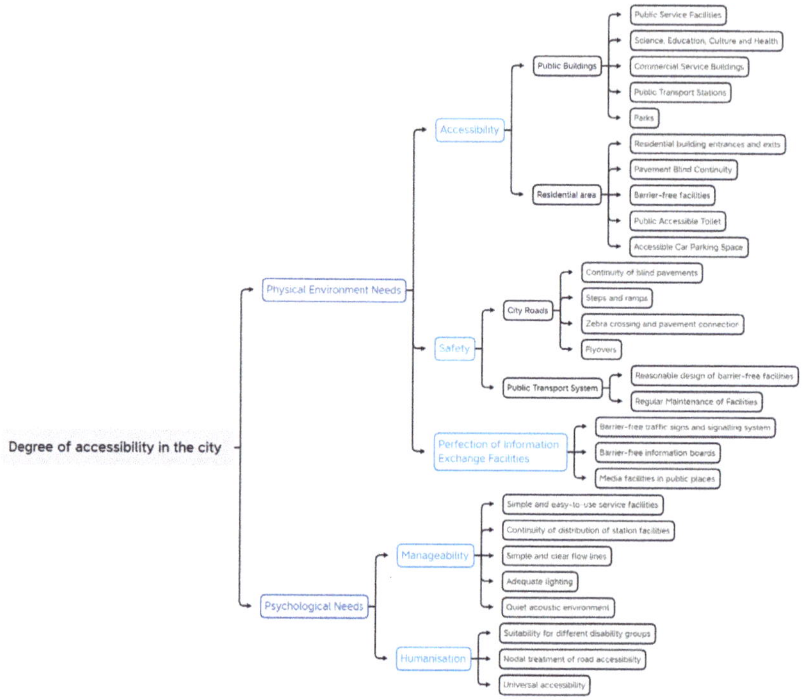

Figure 5. Factors of the evaluation (source: the authors).

In the preceding section, the hierarchical analysis approach was employed to develop an evaluation system for assessing the level of adequacy of urban accessible facilities, taking into consideration the travel and psychological requirements of individuals with disabilities. The evaluation formula was as follows:

$$Final\ Score = \sum\left(\sum(Initial * level3\ factor\ weight) * level2\ factor\ weight\right) * level1\ factor\ weight \quad (1)$$

The specific weight values are shown in the Tables 3–5.

Table 3. Primary evaluation factors.

Classification	Primary Evaluation Factor	Weight
Physical environment needs	Accessibility Security	0.443 0.316
Psychological needs	Controllability Humanisation	0.168 0.072

Table 4. Secondary evaluation factors.

Primary Evaluation Factor	Secondary Evaluation Factor	Weight
Accessibility	Public Buildings	0.413
	Residential areas	0.259
Security	City roads	0.178
	Public transport system	0.094
	Information-exchange facilities	0.056
Controllability		
Humanisation		

Table 5. Tertiary evaluation factors.

Secondary Evaluation Factor	Tertiary Evaluation Factor	Weight
Public buildings	Public service facilities	0.587
	Science, education, culture and health buildings	0.226
	Commercial service buildings	0.100
	Public transport stations	0.051
	Parks and green spaces	0.035
Residential areas	Residential building entrances and exits	0.605
	Continuity of blind pavements	0.214
	Accessibility facilities	0.097
	Public accessible toilets	0.005
	Barrier-free car parking spaces	0.034
City roads	Continuity of blind pavements	0.698
	Steps and ramps	0.186
	Zebra crossing and footpath junction	0.069
	Footbridge	0.046
Public transport system	Reasonable design of accessible facilities	0.800
	Regular maintenance of facilities	0.200
Information-exchange facilities	Barrier-free traffic signs and signalling systems	0.742
	Barrier-free information boards	0.183
	Media facilities in public places	0.075
	Simple and easy-to-use service facilities	0.499
	Continuity of distribution of station facilities	0.285
	Simple and clear flow lines	0.128
	Adequate lighting	0.057
	Quiet acoustic environment	0.032
	Suitability for different disability groups	0.528
	Nodal treatment of road accessibility	0.333
	Universal accessibility	0.140

4. Case Study

Jinan possesses a rich historical background, and the city is continuously undergoing modernization and advancement. The spatial configuration and economic progress of its districts exhibit significant disparities, particularly between the ancient cultural urban areas and the newly developed districts, which feature distinct environments and varying levels of construction quality. Furthermore, the policies governing these districts are subject to frequent updates. Consequently, the research on Jinan's districts encompassed a diverse range of sample types.

4.1. The Progression of Accessible Facilities Construction in Jinan City

Starting from 2006, after more than five years of unremitting efforts, the provincial capital Jinan's barrier-free construction policies, laws, regulations, and standards system was initially established, the number of accessible facilities further increased, the quality was further improved, the layout was further rationalised, the whole society cared more and more about and supported a barrier-free development of the city, and the construction of urban-area accessible facilities saw some results. In September 2009, the Jinan Municipal People's Government issued The Measures for the Administration of Barrier-Free Facilities Construction in Jinan [28]. The measures established uniform provisions on the responsible departments, construction requirements, and construction projects of accessible facilities in Jinan, thus placing the construction of barrier-free facilities at the institutional level based on evidence and rules. The introduction of the regulations guaranteed and directed the significance of the promotion and improvement of accessible facilities construction in downtown Jinan as well as other areas under its jurisdiction. Taking this as a guide, Jinan began to strive to become a national barrier-free city, responding to the call of the Ministry of Housing and Construction, the China Disabled Persons' Federation, and the Committee on the Elderly, and began to create a "National Barrier-Free City" in a systematic manner. During this period, downtown pavements, major attractions, large shopping malls, and other places underwent an unprecedented large-scale barrier-free transformation.

The Municipal Construction Commission has issued a comprehensive repair plan that includes the renovation of 139 public buildings and facilities, such as Walmart, to make them barrier-free. Additionally, 10 parks and scenic spots, including Daming Lake and Baotu, will have barrier-free entrances, ramps, passages, aisles, staircases, and steps installed. Quancheng Square will also have a barrier-free lift platform established. Furthermore, 15 public transport facilities, such as the Remote Wall International Airport and East Railway Station, will undergo renovations based on their specific requirements. Major public places will see the installation of 530 Braille bus stops, while 934 major bus stops in urban areas will have passenger boarding and alighting guardrails implemented. Moreover, major bus stops in urban areas will have 2349 bus shelter seats installed. Lastly, the city's 196 residential neighbourhoods and corresponding residential buildings will be retrofitted with barrier-free pavements, green spaces, and public service facilities. Figure 6 is several real scenes of barrier-free facilities taken during the research.

4.2. Methods of Research

This paper summarises the behavioural characteristics, psychological features and group needs of disabled groups by investigating the awareness of various groups through a literature search and field observations to understand the needs and travelling characteristics of the barrier-free groups, and through field research, combined with interview and questioning methods to investigate the views of the disabled on the accessibility of the district's medium view of the trend of renewal. We explored the construction of accessible facilities in 26 subdistricts within the 5 main city districts of Jinan and analysed the current distribution of accessible facilities in the main city districts of Jinan, including the continuity, uniformity, and perfection of their distribution. Through this accessibility infrastructure research, the public's perception of the overall situation of accessibility in each area was analysed, and a visual analysis of the overall accessibility infrastructure construction in each area of Jinan City was obtained.

4.3. Research Material

This study covered 26 districts in Jinan City, including Guaiyin, Tianqiao, Shizhong, Lixia, and Lixiancheng, as well as five representatives of the central business district. There were also more than 240 research roads, 384 public buildings (including 125 bus stops, 7 large public parking lots, and 5 underground stations), 159 residential neighbourhoods, 17 parks and green spaces, and 11 historical structures. The study's overall mileage was approximately 169 km.

Figure 6. Partial accessibility facilities in the research area (**a**) Accessible ramp. (**b**) Accessible public toilet. (**c**) BRT accessible ramp and blind path. (**d**) Barrier-free elevator. (Source: Author).

4.4. Initial Examination of Research Findings

The prevailing public sentiment centred around the road infrastructure, residential structures, and commercial amenities situated alongside the park. This emphasis highlights the constrained nature of barrier-free construction efforts within the historic district, which is hindered by the need to preserve historical buildings and showcase external attractions. Consequently, the accessibility needs of disadvantaged residents and their corresponding psychological need for safety have been overlooked.

4.5. Assessment of Accessibility Infrastructure in Jinan Municipality

4.5.1. Examination of the Current State of Accessibility Features in Public Structures

The condition of accessible facilities in the three pavilions within the scope of the study was good; road sections such as Dongying Road, Xingfusi North Road, Weihai Road, Qingdao Road, Weifang Road, and other sections in Huaiyin District achieved a full coverage of blind alleys, and the setting of ramp edges was relatively perfect; recreational public buildings such as Wanshiang City, Evergrande Fortune Centre, Impression Jinan-Quanworld, Hongyang Plaza of Jinan, and Shandong Provincial Congress Theatre achieved a full coverage of barrier-free entrances and exits; and public buildings such as Hengyuan School, BeiduoYi Kindergarten, the Second Experimental Kindergarten in

Huaiyin District, Hengxin Primary School, and Majiyuan Kindergarten in Huaiyin District, and other educational public buildings also set up barrier-free entrances and exits. Other key facilities in other districts, including seven Tianqiao districts, three Shizhong districts, six Lixia districts, and four Licheng districts, also showed a generally good progress in terms of barrier-free construction. Many public places in Jinan were equipped with accessible facilities to protect the rights and interests of the disabled, such as the construction of ramps to ensure the smooth passage of prams and wheelchairs, the construction of barrier-free toilets in some parks to ensure the normal physiological needs of the disabled, the addition of barrier-free access at bus stops, and the setting up of guided tours for the blind at intersections to meet the needs of the blind.

This study analysed three distinct areas, namely the Quancheng Square area, the Quancheng Park area, and the Old Shangfu area, in order to identify the distribution of problematic points related to accessible facilities within each subarea. The analysis focused on various urban elements such as roads, bus stations, residential buildings, parks, green spaces, and buildings with historical and cultural significance. The aim was to provide a refined and systematic understanding of the identified issues, which is visualized in Figure 7.

Figure 7. Distribution of problematic accessibility points and sections in the old town area (source: the authors).

Through the picture analysis, the continuity of the roads on the south and north sides of the park belt was good; however, the perfection and distribution were not good enough, which was reflected in the better degree of perfection of the barrier-free facilities on the south side of the base. A comparison with the buffer zone showed that there were localised areas of missing barrier-free facilities within 300–500 m of the buffer zone for public service facilities. In older neighbourhoods, such as the public service facilities around the settlement of the city of Zakyn, the accessibility index was poor and there were many missing locations, while the ageing phenomenon was serious, and the proportion of elderly residents with weak mobility was even higher, which required a more complete configuration of barrier-free facilities.

In the context of privately owned communities and real estate, as well as distant places like the southern highland areas, there persisted a prevalent deficiency in accessibility and inconsistencies in the development of accessible infrastructure.

4.5.2. Examination of Visual Impairment Continuity

Through research interviews with several handicapped individuals, it was revealed that the degree of continuity of accessible roads was a very major hurdle to disabled people's travelling. The resulting bar chart is shown below in Figure 8.

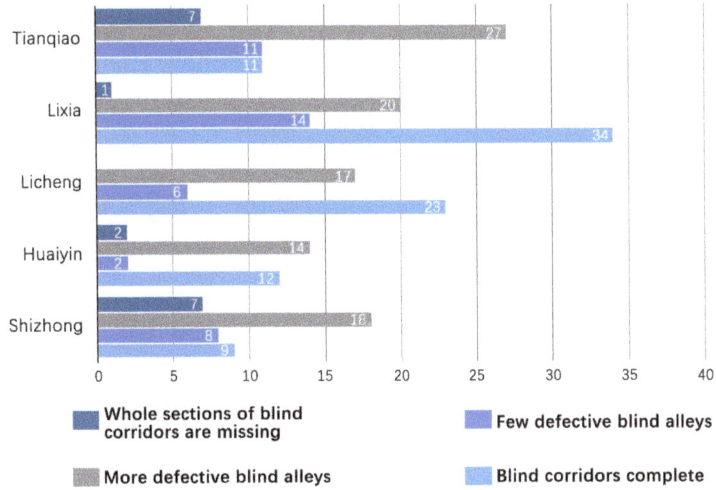

Figure 8. Road accessibility analysis (source: the authors).

Statistics on road facilities for the blind in each sub-district of the five districts in the main urban area of Jinan City found that the overall situation of the integrity of the blind on the roads accounted for a small percentage of the total roads researched, as each district was below 35%. Figure 9 shows the situation of blind access facilities in different areas on a map.

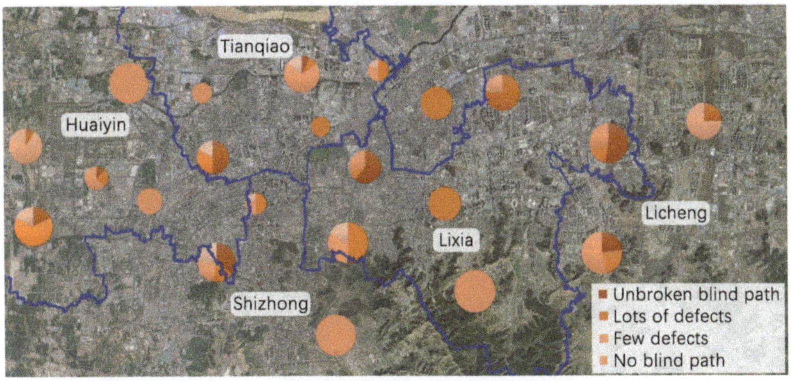

Figure 9. Summary map of Jinan Road facilities' accessibility analysis (Source: Author).

4.5.3. Examination of Accessible Facilities in Residential Structures

We conducted an in-depth investigation into the state of accessible infrastructure in residential buildings, including the examination of facilities such as lift shafts and public places within the community.

The distribution of accessible facilities in residential buildings exhibited a pattern similar to that observed in road transport facilities (Figure 10). Specifically, Lixia District demonstrated the most favourable overall situation, which was roughly commensurate with the level of economic development in each respective region. However, there were relatively fewer areas that boasted superior overall accessible facilities, indicating the need for a further improvement in the overall barrier-free environment.

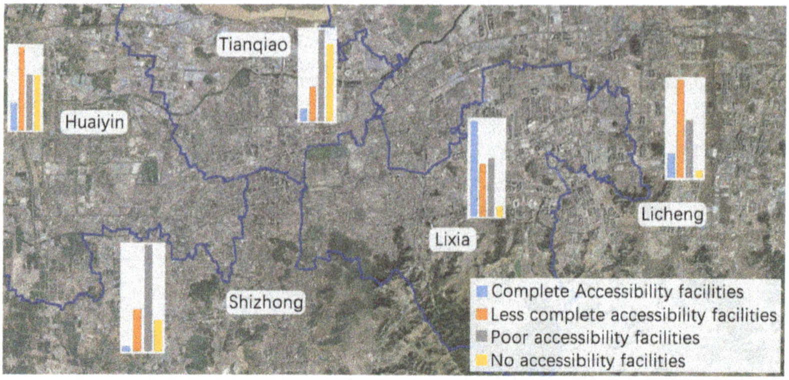

Figure 10. Summary map of analysing accessibility in residential neighbourhoods (source: the authors).

4.5.4. Examination of Accessible Facilities in Environmentally Friendly Parks

The parks were divided into seven gradients using a gradient division method, which was based on the development of barrier-free services in green areas. To visually represent this division, a map was utilised with a visualisation technique.

A visual analysis showed in Figure 11 that the accessibility of parks and green spaces in Jinan City was not good enough. According to this analysis, it could be seen that the areas with better accessible facilities were concentrated in the central area of Jinan City, and the accessible facilities in the old city areas were improved based on the existing ones, such as the Gucheng, Honglou, and Tianqiao areas, while the new city areas, such as the Hanyu and Long'ao areas, had a low degree of improvement of accessible facilities; in addition, the remote areas not part of the centre of the urban area of Jinan City, such as the Longdong area, were generally free of barriers, and the southern mountainous areas had a low degree of improvement. From this, it could be seen that the coverage of barrier-free facilities in Jinan City was not wide enough, and the areas with perfect barrier-free facilities were only limited to the central urban area, historical preservation areas, and areas with a better economic development, while the coverage and perfection of remote areas were insufficient. Figure 12 shows some areas where the accessibility facilities are incomplete.

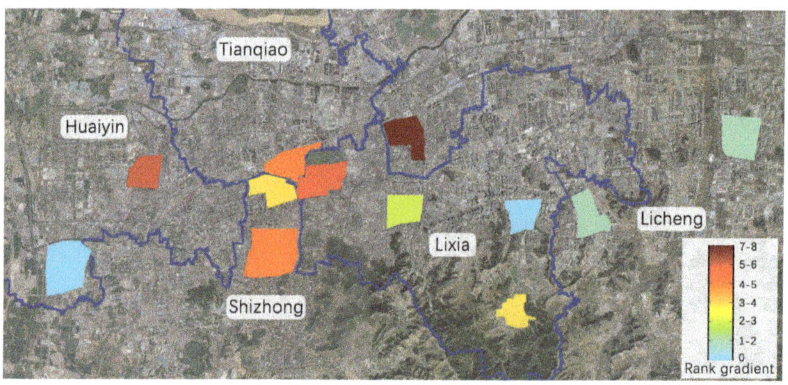

Figure 11. Summary chart of the analysis of the accessibility improvement of parks and green spaces (source: the authors).

Figure 12. Some inadequate accessibility facilities. (**a**) There are still no ramps in some small parks and there is a large height difference. (**b**) The blind path ended abruptly (source: the authors).

4.6. Assessment of Accessible Infrastructure in Researched Environments

Based on the findings of the study, the evaluation of accessible facilities in Huaiyin, Tianqiao, Shizhong, Lixia and Licheng was conducted utilising the assessment framework outlined in the preceding section whose results are arranged in Figure 13.

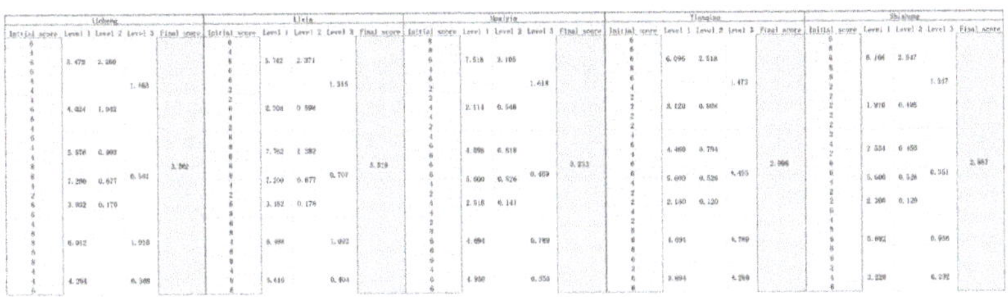

Figure 13. Score by region (source: the authors).

The districts' scores, arranged in descending order, were as follows: Lixia: 3.518, Lixing: 3.362, Huaiyin: 3.233, Tianqiao: 2.998, Shizhong: 2.887. The accessibility was the highest in the locality of Huaiyin, and the safety, controllability, and humanisation were all the highest in Lixia. Based on the evaluation results, the corresponding factors will be updated or improved.

5. Conclusions and Recommendations

This study presents conclusions and recommendations regarding the existing issues surrounding accessible facilities. The recommendations primarily emphasise two aspects. On the one hand, it is to propose improvement strategies for the accessibility of Jinan's researched area based on the results of the research and evaluation, and on the other hand, it is to make suggestions for an inclusive urban renewal based on the factor weights in the evaluation system.

5.1. Strategies for Increasing Accessibility in Selected Jinan Districts Based on Evaluation Results

Bar charts as shown in Figure 14 were made to visually compare the district-level factor-weighted scores , and generalised improvement strategies were proposed for each of the five research areas.

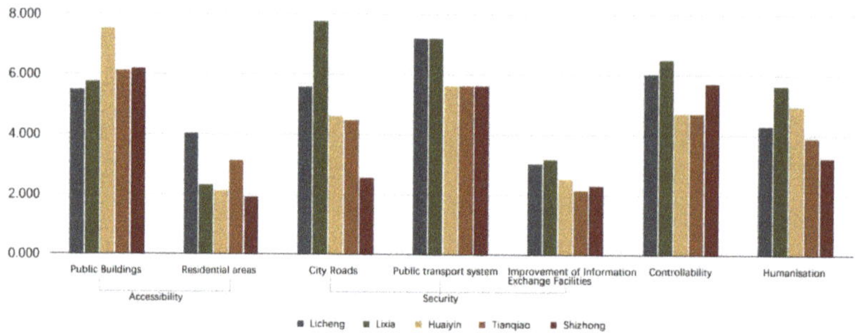

Figure 14. Comparison of level 1 scores of the five districts (source: the authors).

The five districts were all poor in terms of accessibility, information-exchange facilities, and humanisation of residential areas, and should be updated with accessible services such as bus stops, traffic signal systems, and accessible nodes, as well as maintaining blind alleys at entrances and exits of buildings and pavements in residential areas. The continuity of urban roads in the central district was also poor, and blind facilities, pavements, and zebra crossings should be managed and corrected to improve the continuity of roads.

5.2. Urban Renewal Recommendations Based on Evaluation Systems

Based on the evaluation scores, recommendations were made for the renewal of barrier-free facilities: priority was given to the renovation of barrier-free facilities in public buildings and the planning of barrier-free environments in residential areas, followed by the improvement of the accessibility of urban roads and the accessibility of the public transport system, and lastly, the management and maintenance of small-scale barrier-free service facilities, and so on.

1. New public buildings should strictly comply with barrier-free design norms and build barrier-free facilities with high quality; old public buildings with substandard facilities should be rectified and supervised within a limited period and follow-up maintenance management should be strengthened [29];
2. In residential areas, the barrier-free facilities that should be paid attention to include ramps, passages, doors, handrails and lifts, etc.; road accessible facilities should be improved with kerb ramps, blind corridors, etc.; and ancillary public service facilities, such as public toilets, accessible parking spaces, low-level service facilities and lifting platforms, etc., should also be fully equipped [30];
3. The improvement of road continuity should focus on the management and maintenance of blind alleys, connecting height differences with ramps, etc.; the standardisation strategy of the transport system includes: regulating the taxi industry to eliminate the phenomenon of taxis refusing to carry the disabled, transforming bus stop signs set in a high position and with a small font, and in terms of traffic management, increasing the punishment of vehicles occupying blind alleys, etc. [31];
4. Other accessible service facilities should be more user-friendly, easier to use, and regularly maintained and managed, and nonstandard facilities should be remedied to reduce inaccessibility or poor accessibility [32];
5. With the development of information technology and intelligence, intelligent assistive facilities for disabled people travelling have been paid attention to and researched, and various intelligent accessible facilities such as smart blind canes, smartphones with accessibility functions, and smart wheelchairs have appeared one after the other; some scholars have also researched the monitoring of high-risk zones for vulnerable groups, such as the elderly, who are prone to fall in outdoor environments [33], which provides a scientific basis for the construction of the public service field and enhances the convenience and safety of disabled people's travels as well as the efficiency and accuracy of the improvement of the barrier-free environment.

6. Conclusions and Prospect

At present, Chinese cities are in the stage of quality improvement, and there is still room for exploring research on urban renewal. The evaluation system established in this paper is only a tool and basis for one branch of this renewal. On this basis, more in-depth and comprehensive research will be carried out at a later stage to gradually build more inclusive and livable cities.

Author Contributions: Methodology, Y.W.; software, Y.Y. and J.W.; validation, Y.W. and Y.L.; formal analysis, Y.W. and Y.Y.; investigation, Y.L.; writing—original draft preparation, Y.W.; writing—review and editing, Y.Y. and J.W. All authors have read and agreed to the published version of the manuscript.

Funding: This research received no external funding.

Data Availability Statement: All data used in this research are available on open data sources.

Conflicts of Interest: The authors declare no conflict of interest.

References

1. Wang, P. *Wang Pingping: The Total Population Decreased Slightly and the Level of Urbanization Continued to Improve;* State Statistics Bureau: Beijing, China, 2023.
2. 2030 Agenda for Sustainable Development. New York, USA, 2016; Available online: https://sdgs.un.org/2030agenda (accessed on 20 October 2023)
3. National Development and Reform Commission. *Outline of the 14th Five-Year Plan (2021–2025) for National Economic and Social Development and Vision 2035 of the People's Republic of China;* National Development and Reform Commission: Beijing, China, **2021**.
4. Huang, B.; Lv, X.; Fang, Y. Survey and Reflection on the Status of accessibility facilities. *For. Sci. Technol. Intell.* **2023**, *55*, 165–167.
5. He, X.; Yang, Y. A Review of the Current Research Status of Accessible Design Methods. *Packag. Eng.* **2023**, *44*, 60–73. [CrossRef]
6. Boix-Cots, D.; Pardo-Bosch, F.; Pujadas, P. Analysis and Comparison of the Infrastructure Report Cards as a Decision Making Tool for Sustainable Development. *Buildings* **2023**, *13*, 2166. [CrossRef]
7. Huang, X.; White, M.; Langenheim, N. Towards an Inclusive Walking Community—A Multi-Criteria Digital Evaluation Approach to Facilitate Accessible Journeys. *Buildings* **2022**, *12*, 1191. [CrossRef]
8. Tang, Y.; Liu, Y.; Wang, Y. From Growth to Inclusion: The Multidimensional Connotation and Assessment of Inclusive Cities from the Perspective of the Right to the City. *J. Shanghai Adm. Coll.* **2023**, 48–64.
9. Harris, E.; Franz, A.; O'Hara, S. Promoting Social Equity and Building Resilience through Value-Inclusive Design. *Buildings* **2023**, *13*, 2081. [CrossRef]
10. United Nations. New Urban Agenda. In Proceedings of the Section III United Nations Conference on Housing and Sustainable Urban Development, Quito, Ecuador, 20 October 2016.
11. Asian Development Bank. *Strategy 2020: The Long-Term Strategic Framework of the Asian Development Bank 2008–2020;* Asian Development Bank: Manila, Philippines, 2008.
12. Xiang, D. Implications of the Concept of Inclusive Development for the Construction of Social Policy in China. *Soc. Sci.* **2012**, *1*, 70–74.
13. Xu, J. Multi-dimensional analysis and visualisation of inclusive design methodology. *Design* **2021**, *34*, 102–105.
14. Jing, X.; Xia, J.; Chen, H.; Wang, X. A study on the low utilisation rate of barrier-free facilities from the perspective of disabled people—A case study of Nanjing city. *Urban Plan.* **2020**, *44*, 47–56.
15. Zhao, P.; Chen, J. Study on the continuity of barrier-free design at the boundary of public space in old urban areas–Hangzhou City as an example. *Intell. Build. Smart City* **2022**, *8*, 160–164. [CrossRef]
16. Yuan, Z.; Liu, T.; Shao, L. Accessibility Needs and Satisfaction of the Physically Disabled in Beijing's Quarter-hour Community Service Circles. *Planner* **2019**, *35*, 25–31.
17. Wu, Y.; Gao, G.; Wu, M.; Wang, X.; Xue, P.; Gou, B. Investigation and analysis of the application status of urban barrier-free facilities (blind passage). *Chin. Tissue Eng. Res.* **2020**, *24*, 271–275.
18. Zhang, M.; Fu, J.; Liu, J.; Yan, M. Research on Urban Public Transport Travelling Service for Disabled People. *Packag. Eng.* **2022**, *43*, 199–207. [CrossRef]
19. Andersen, A.D.; Andersen, P.D. Foresighting for inclusive development. *Technol. Forecast. Soc. Chang.* **2017**, *119*, 227–236. [CrossRef]
20. Hinrichs, M.M.; Johnston, E.W. The creation of inclusive governance infrastructures through participatory agenda-setting. *Eur. J. Futures Res.* **2020**, *8*, 15. [CrossRef]
21. Gillovic, B.; McIntosh, A. Accessibility and Inclusive Tourism Development: Current State and Future Agenda. *Sustainability* **2020**, *12*, 15. [CrossRef]
22. Sun, J. *A Pattern Language-based Participatory Planning Approach for Barrier-free Communities;* Harbin Institute of Technology: Harbin, China, 2021.
23. Maslow, A.H. *Motivation and Personality;* Harper and Row: New York, NY, USA, 1970.

24. Hong, C.; Donghui, Z. Barrier-free design of residential areas and characteristics of disabled people. *Sichuan Archit. Sci. Res.* **2009**, *35*, 259–261.
25. Jin, C. A Study on Barrier-Free Travelling for People with Disabilities from a Systems Perspective. Master's Thesis, Guangxi Normal University, Guilin, China, 2019.
26. Cai, Y. Research Review on the Mobility Psychology of Disabled People and Suggestions on Urban Transport Planning. In Proceedings of the Annual Conference on Urban Planning in China 2016, Shenyang, China, 24–27 September 2016; p. 12.
27. Karlqvist, A. *Some Theoretical Aspects of Accessibility-Based Models//Dynamic Allocation of Urban Traffic Models*; Columbia University Press: New York, NY, USA, 1971.
28. General Office of Jinan Municipal People's Government .*Jinan City Barrier-Free Facilities Construction Management Measures*; General Office of Jinan Municipal People's Government: Jinan, China, 2009; pp. 8–9.
29. Sun, Z. A Study on Enhancing the Satisfaction of Barrier-Free Facilities Construction. Master's Thesis, Tianjin University, Tianjin, China, 2021.
30. Zhang, W. Standardised Construction Practice of Barrier-Free Facilities in Urban Residential Areas. *China Stand.* **2021**, *2*, 145–149.
31. Ye, H. Jinan: Barrier-Free Facilities for the Disabled Are Difficult to Satisfy. *China Disabl. People* **2013**, *12*, 6.
32. Zhang, S.; Xu, B. Barrier-free facilities still face the "three no-uses" barrier. *Beijing Dly.* **2023**, *8*.
33. Meng, J.; Yang, L.; Lei, H. Promoting Elderly Care Sustainability by Smart Village Facilities Integration—Construction of a Public Service Field with Introduction of Fall Posture Monitoring. *Buildings* **2023**, *13*, 2144. [CrossRef]

buildings

MDPI

Article

Field Investigation of Water Infiltration into a Three-Layer Capillary Barrier Landfill Cover System Using Local Soils and Construction Waste

Yuedong Wu [1,2], Jincheng Ren [1,2] and Jian Liu [1,2,3,*]

1 Key Laboratory of Ministry of Education for Geomechanics and Embankment Engineering, Hohai University, Nanjing 210098, China; hhuwyd@163.com (Y.W.); 211604010056@hhu.edu.cn (J.R.)
2 Geotechnical Engineering Research Center of Jiangsu Province, Nanjing 210098, China
3 Engineering Research Center of Dredging Technology of Ministry of Education, Hohai University, Changzhou 213000, China
* Correspondence: 20170053@hhu.edu.cn; Tel.: +86-1836-296-0323

Abstract: In response to the rapid urban expansion and the burgeoning number of landfill sites, managing water infiltration in these areas has become a critical challenge, especially in cities like Shenzhen, Hong Kong, and Singapore where traditional cover materials such as silt, clayey gravel, and sand are scarce. A three-layer (silt/gravelly sand/clay) capillary barrier cover system has been proposed to address this issue in humid climates. As an alternative to scarce traditional materials, using local soils and construction waste (CW) for this system presents a viable solution. However, the real-world performance of this adapted three-layer system, constructed with local soils and CW under natural rainfall conditions, remains to be fully evaluated. This paper presents a field test evaluating the water infiltration behavior of a three-layer capillary barrier landfill cover system under natural conditions. The tri-layered system is comprised of a 0.6 m loose local unscreened soil layer, covered by a 0.4 m CW layer and topped by a 0.8 m heavily compacted local screened soil layer. Monitoring findings reveal that, during the wet season, infiltration through the top two layers was staved off until the third rainfall, after which these layers retained moisture until 15 September 2016. The fluctuation in pore water pressure in the topmost layers showed each rainfall was contingent not only on the day's precipitation but also the hydraulic state. Beyond the hydraulic state's influence, a deeper tensiometer showed resulted in a diminished correlation between the surge in pore water pressure and daily rainfall. This declining correlation with depth can be attributed to the capillary effect and the reduced permeability of the screened soil layer. Rainfall patterns significantly affect percolation, with the combination of a short-duration, intense rainfall and prolonged weak rainfall resulting in a marked increase in percolation. In the foundational screened soil layer, the pore water pressure remained relatively low, with the cumulative percolation over six months (June to December) registering approximately 10 mm. These findings suggest a promising performance of the three-layer capillary barrier cover system, integrating local soils and CW, in the year of the study conducted in a humid environment.

Keywords: capillary barrier; infiltration; field test; local soil; construction waste

1. Introduction

Recently, a three-layer capillary barrier cover system has been proposed to minimize water infiltration into municipal solid waste in humid climates [1,2]. The three-layer cover system consists of a fine-grained soil layer (e.g., silt layer) overlying a coarse-grained layer (e.g., gravelly sand layer), which in turn overlies a fine-grained soil layer such as clay layer. Results obtained from numerical simulations [1], soil column tests [3], and physical flume model tests [2] show that the three-layer cover system performed well under heavy rainfalls. The breakthrough of the upper two layers occurred after heavy rainfalls. The percolation through the three-layer capillary barrier cover system was negligible, primarily because the low permeability of the bottom clay layer redirected most of the infiltrated rainwater laterally along the gravelly sand layer. This transformation meant that the gravelly sand layer's role shifted from a capillary barrier to a lateral diversion passage, due to the influence of the clay layer following the upper layers. Recently, construction waste has been extensively used in engineering construction. An optimized system based on the three-layer capillary barrier cover has been proposed [4], particularly addressing the challenges of material availability in urban areas such as Shenzhen, Hong Kong, and Singapore. To reduce costs in a landfill cover system, construction waste was used instead of gravel in the middle layer, and local soil replaced kaolinite. Additionally, a vegetative layer was added on top. A field test was conducted to assess the system's impermeability. However, the performances of this three-layer capillary barrier system in responding to different rainfall patterns in field tests are limited.

Field monitoring programs have been carried out extensively on traditional landfill covers, for instance, compacted clay cover [5], evapotranspiration cover [6,7], two-layer capillary barrier [8,9], and multi-layered cover [10]. In these field sites, monitoring data were usually recorded by a data logger or collected manually at a low frequency. However, some rainfalls only lasted a few hours and thus the data may be missed. Moreover, landfills usually are located far away from downtown and thus lack computers and electricity for the data logger. To solve these difficulties, a wireless system based on a General Packet Radio Service (GPRS) network was used to establish the monitoring system at a high frequency [11]. In this system, the computer could be fixed far away from the testing site (e.g., landfill) and the instruments can be installed in the testing site where the GPRS network covers. However, the GPRS network may not be stable sometimes due to the interference from significant noise and impairments at the site [12]. Under such conditions, the monitoring system in the landfill may not work effectively.

In this study, an architecture with local and GPRS networks at a high logging frequency was established in the landfill where the GPRS network may not be good sometimes. Based on this architecture, the study investigated the performance of a three-layer cover system using local soils and construction waste (CW) under various natural rainfall conditions in a humid climate. Additionally, the research explored the performances of the three-layer capillary barrier system responding to different types of rainfall. For each heavy rainfall, the responses of measured pore water pressure in the local soils and CW were measured. In addition, the percolation was directly measured with drain gauges to better understand the performance of the three-layer cover system using the local soils and CW.

2. Materials and Methods

2.1. Testing Site

The testing site is located at the Xiaping solid waste landfill in Shenzhen, Guangdong China, as shown in Figure 1. The Xiaping landfill, the largest one in Shenzhen, receives an average of 5300 t of domestic waste per day. More than 6 Mt domestic wastes have been dumped in the landfill. After the landfill reaches the designed height, a final cover system has been installed over the waste.

This testing site is in a humid region with an average annual rainfall of about 1932.9 mm. The daily evaporation potential values are about 4.6 and 3.3 mm in wet and dry seasons, respectively. The average monthly maximum and minimum temperatures range from 19.8 °C to 32.3 °C and from 12.5 °C to 26.3 °C, respectively. No freezing temperature was observed at the site.

An area, 6 m wide by 20 m long, was selected on a slope along the western side of the Xiaping solid waste landfill. The test area has an average slope angle of about 15° and a height of 5 m. Based on the conceptual framework from Ng et al. [1,2], the three-layer capillary barrier cover system in this study consisted of a 0.6 m thick loose local unscreened soil layer overlying a 0.4 m thick construction waste layer and a 0.8 m thick heavily local compacted screened soil layer. The screened local soil was obtained by screening the local soil through a layer of metal mesh with a 10 mm aperture.

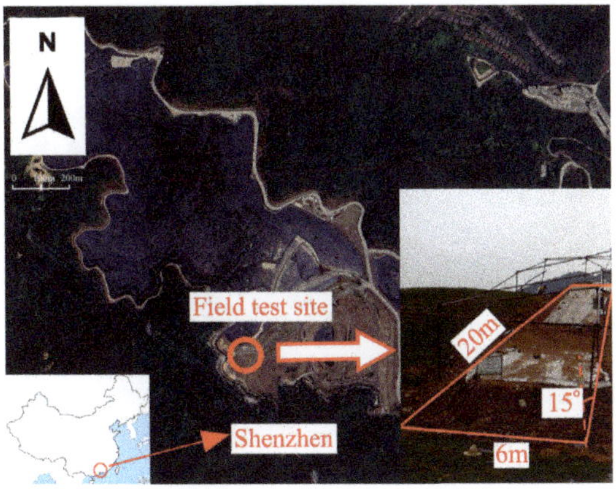

Figure 1. Overview of the field test site.

The cover system primarily consists of three layers: unscreened soil, construction waste, and screened soil. The unscreened and screened soils are sourced locally, while the construction waste predominantly comprises waste concrete and bricks. These types of construction waste are generally considered to be environmentally inert. This implies that under natural conditions, they are less prone to chemical reactions.

2.2. Materials and Construction Methods

A sampling program was conducted during the construction of the cover. The samples were analyzed for particle size distribution (ASTM D 422) [13], Atterberg limits (ASTM D 4318) [14], compaction characteristics (ASTM D 698) [15], and saturated permeability properties (ASTM D 5084) [16]. Table 1 summaries the physical properties of the three soils (i.e., unscreened soil, CW, and screened soil). The unscreened soil exhibited D60, D30, and D10 values of 0.298 mm, 0.091 mm, and 0.006 mm, respectively, while the screened soil showed values of 0.165 mm, 0.032 mm, and 0.003 mm. The fine content was 25.2% for the unscreened soil and 40% for the screened soil. After screening, an increase in fine content was observed, leading to a decrease in the saturated hydraulic conductivity (k_s). The main difference between the CW layer and the local soil lies in the overall larger particle size, with D60, D30, and D10 values of 20.1 mm, 8.9 mm, and 2.3 mm, respectively. The saturated hydraulic conductivity of the CW layer is 6–7 orders of magnitude higher than that of the screened soil, enabling it to form a cover system with capillary barrier effects (CCBE) with the unscreened soil. Construction waste mainly consists of inert waste, non-inert waste, and contaminated waste; the CW chosen for this study mainly consists of recycled crushed concrete and bricks. The particle size of these materials determines their hydraulic properties. These are classified as inert waste and have no environmental impacts under natural conditions [17,18]. For the upper two layers, the relative densities of unscreened soil and CW were 80% and 94%. For the bottom layer, the screened soil was compacted to 95% of its maximum dry density, aiming to obtain a low coefficient of water saturated permeability and porosity.

The three-layer landfill cover was constructed from July 2015 to May 2016. The soils were compacted at four sublayers for the bottom layer, three sublayers for the middle layer, and three sublayers for the top layer using a mini roller compactor with a weight of 1 t. As suggested by Ng et al. [2], geotextiles were installed at the interface between the two soil layers at depths of 0.6 and 1.0 m to prevent soil particle migration. The geotextiles used in our study were non-woven, with a thickness of 0.9 mm, and had an equivalent pore size ranging from approximately 0.07 mm to 0.2 mm, the saturated permeability of geotextile (i.e., 7×10^{-2} m/s) was significantly higher than that of the uppermost soil layer (i.e., $5 \times 10^{-7} \sim 9 \times 10^{-7}$ m/s), water penetration was not affected. After the construction of the cover system, the instrumentation and monitoring system were installed at the testing site.

Table 1. Properties of local soils and construction waste used in the field test.

Soil Type		Unscreened Soil	Construction Waste	Screened Soil
Unified soil classification system		SC	GW	SC
Liquid limit, *LL* (%)		37	/	37
Plastic limit, *PL* (%)		20	/	20
Grain size distribution	D_{60} (mm)	0.298	20.1	0.165
	D_{30} (mm)	0.091	8.9	0.032
	D_{10} (mm)	0.006	2.3	0.003
	Fine content (0.075 mm; %)	25.2	/	40.0
	Coefficient of uniformity, C_u	49.6	8.7	55
	Coefficient of curvature, C_c	4.6	1.7	2.1
Maximum dry density, ρ_d (Mg/m^3)		1.86	1.89	1.82
Saturated coefficient of permeability, k_s (m/s)		5×10^{-7}–9×10^{-7}	6×10^{-2}–1×10^{-1}	1.0×10^{-8}–3.0×10^{-8}

2.3. Instrumentation Program

2.3.1. Layout of Instruments

Figure 2 shows the layout of all instruments used in the field monitoring program. The instruments included a rain gauge for recording precipitation, jet-filled tensiometers with pore pressure transducer for negative pore water pressure, and drain gauges for percolation. The details for each instrument are summarized in Table 2.

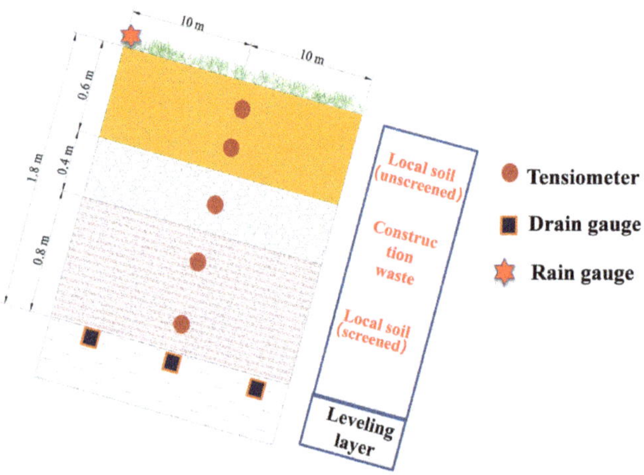

Figure 2. Layout of instruments in the landfill cover.

Five tensiometers were installed only at the middle section of the cover system. These tensiometers were spaced 0.5 m apart. For each depth (i.e., 0.2 m, 0.4 m, 0.8 m, 1.2 m, and 1.6 m), there was one tensiometer to study water seepage characteristics in the three-layer cover system. The rain gauge was installed at the crest of the slope. These instruments were connected to a wireless network system for data collection. For detailed information on the Data Transmission System, please refer to the Appendix A.

Table 2. Summary of instruments.

No.	Measurement	Type of Instrument	Quantity	Measuring Range/Accuracy
1	Pore water pressure	Jet-fill tensiometers	5	−90–100 kPa/0.1 kPa
2		Pore pressure transducers	5	
3	Percolation	Drain gauge	3	No limit/0.1 mm
4	Precipitation	Rain gauge	1	0–50 mm/min/0.2 mm/min

2.3.2. Instruments, Calibrations, and Installing Methods

Rain gauge—The rain gauge is a freestanding receptacle for measuring precipitation. As water is collected, the tipping bucket fills to the point that it tips over, causing a momentary closure of a switch to incrementally measure rainfall accumulation. This action empties the bucket in preparation for additional measurement. Water discharged by the tipping bucket passes out of the rain gauge with no need for emptying. A simple device consisting of a peristaltic pump with transparent hose was used for calibration as presented by Vasvári [19].

Jet-filled tensiometer—Negative pore water pressure was measured by jet-filled tensiometer manufactured by the Soil moisture Equipment Corporation [20]. In order to increase the accuracy and automatically record data, a pore pressure transducer (PPT) was employed, taking the place of a vacuum gauge. Different from the traditional artificial reading for the vacuum gauge, the use of PPT could decrease the workload of reading and increase the recording frequency to meet the requirements for recording water infiltration during the process of rainfall. Before installation, all tensiometers with PPT were calibrated using the method proposed by Chen et al. [21] and Feng et al. [22]. The measurement accuracy of PPT was 0.1 kPa, about 5% of the accuracy of vacuum gauge.

Drain gauge—The drain gauge was a buried tank with an open top that collected any infiltrated water from the overlying soil mass. The diameter of the tank was designed as 1 m. In the tank, the soil mass was compacted with a water content to a dry density identical to the surrounding soil mass. At the bottom of the tank, a drain tube with porous stone was connected to carry water from the tank to a collection tank. The mass of the infiltrated water collected in the collection tank was measured by a balance. Then, the infiltration rate can be calculated by the mass changes of the collection tank over the measuring duration. The accuracy of mass measurements is 0.5 g. it has been widely adopted and confirmed to exhibit outstanding reliability and high precision [23].

3. Results

3.1. Pore Water Pressure

Figure 3 shows the record of daily precipitation from 1 June 2016 to 1 December 2016. After construction, the cover system was first subjected to a wet season lasting about five months and then a dry season lasting about 1 month. According to data from the Shenzhen Meteorological Bureau [24], the total precipitation in Shenzhen for the year 2016 was 2490.6 mm. The maximum and minimum annual rainfalls recorded over the past 50 years in Shenzhen were 2747.3 mm and 1269.7 mm, respectively, indicating that the 2016 precipitation level is representative of the typical climatic conditions in Shenzhen. During the wet season, the condition was a relatively high humidity and the total precipitation was 1551 mm, approximately 65% of Shenzhen's annual rainfall. According to the meteorological report formulated by the Shenzhen Meteorological Bureau [24], the number of rainy days in the rainy season is 64, which accounts for about 45% of the total number of days in this period. To study the effectiveness of the cover system under natural rainfall conditions over a year, daily rainfall data were selected as the metric. It should be noted that eight heavy rainfalls (or rainstorms) whose daily precipitation (labeled in the figure) was higher than 50 mm [25] occurred during the wet season. The maximum daily precipitation was as large as 217.8 mm (No. 7). In the dry season, the soil humidity was low and the total precipitation was only 162.3 mm, with the heavy rainfall event No. 9 contributing 52.2 mm to this total. In the dry season, the number of no-rainfall days reached 23 and the number of the maximum continuous no-rainfall days was 11.

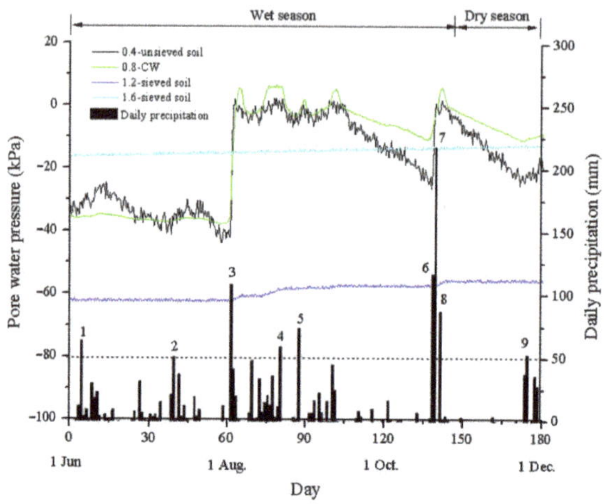

Figure 3. Changes of pore water pressure in the three-layer cover system at different soil depths under natural conditions.

The development of pore water pressure at each depth was also included in Figure 3. Due to the evaporation, the cavitation in the tensiometer at a depth of 0.2 m occurred and its data were not shown in the figure. As expected, the pore water pressure increased in response to each rain at depths of 0.4 and 0.8 m, particularly subjected to nine heavy rainfalls. At a depth of 1.2 m, the increases in the pore water pressure were smaller than those at depths of 0.4 m and 0.8 m. It indicates that little water infiltrated into the screened layer at the bottom [2]. On the other hand, the pore water pressure in the upper two layers decreased in the no-rainfall days, e.g., from 15 September 2016 to 16 October 2016 and from 22 October 2016 to 21 November 2016. In the upper two layers, the pore water pressure in the top layer decreased more than in the layer beneath during periods without rainfall. At depths of 1.2 m, the pore water pressure was maintained as constant. It was due to the low permeability in the CW layer during drying and thus the water loss in the underlying screened soil layers was small. Similar changes in pore water pressure were obtained using numerical simulations by Ng et al. [26]. In this study, "breakthrough" refers to the movement of water through the top two soil layers under rainfall conditions. In this three-layer system, when a breakthrough occurs in the top two layers, the presence of a low-permeability soil layer impedes the infiltration of rainwater. A breakthrough is characterized by pore water pressure, correlating with field capacity. Correspondingly, pore water pressure shows increased sensitivity during rainfall events as soil approaches saturation [27]. It should be noted that before 2 August 2016, the total precipitation was 423 mm, approximately 21% of Shenzhen's annual rainfall. However, the pore water pressure remained relatively low (ranging from −25 to −45 kPa) at the depths of 0.4 and 0.8 m. The breakthrough of the upper two layers did not occur. It may be because the maximum daily precipitation was 63.8 mm, smaller than the rain intensity when the breakthrough of the upper two layers occurred (e.g., No. 3 with 108.8 mm, No. 6 with

116.3 mm, and No. 7 with 217.8 mm). At a depth of 1.2 m, the pore water pressure remained at −62 kPa. After the breakthrough (e.g., No. 3 rainfall with 108.8 mm), the pore water pressure remained near zero until 15 September 2016. In this period, positive pore water pressure was observed in the upper two layers subjected to each heavy rainfall.

The Xiaping solid waste landfill generates landfill gas (LFG) primarily through the biodegradation of municipal solid waste (MSW). The gas composition predominantly consists of approximately 60% methane (CH_4) and 40% carbon dioxide (CO_2). The LFG production rate at this landfill is approximately 1.109×10^3 m^3/h, based on estimates [28]. At a depth of 1.6 m in the screened soil layer, the pore water pressure varied from −10 kPa to −22 kPa, unrelated to the daily precipitation. Its value was approximately 40 kPa larger than that at a depth of 1.2 m in the same soil layer. It may be induced by the high humidity of landfill gas under the cover [29]. In other words, wet landfill gas may be important for a minimum of landfill gas emission [26].

As previously mentioned, the heavy rainfalls were important for the breakthrough of the upper two layers. Figures 4–6 summarize the change in pore water pressure subjected to the nine heavy rainfalls at a depth of 0.4 m (unscreened soil layer), 0.8 m (CW layer), and 1.2 m (screened soil layer). At a depth of 0.4 m, the maximum increase in the pore water pressure was 42.2 kPa subjected to a No. 3 heavy rainfall. However, the daily precipitation of No. 3 rain was 108.8 mm (not maximum). The daily precipitation seemed not to be the sole factor to the change in pore water pressure. It may be due to the hydraulic state of the cover system. After the rains of No. 3, 7, and 8, the unscreened soil was wet. In the wet state, it was easy to reach the maximum pore water pressure for the cover system. The maximum pore water pressure was approximately 2 kPa at a depth of 0.4 m. It should be noted that the maximum pore water pressure was smaller than its hydrostatic pressure (i.e., 4 kPa for the tensiometers at a depth of 0.4 m) by assuming that the cover system was saturated. It may be due to desiccation cracking near the top of the unscreened soil layer [5]. A similar change to pore water pressure also occurred at a depth of 0.8 m. The maximum pore water pressure of 5.5 kPa also occurred after No. 3 rain.

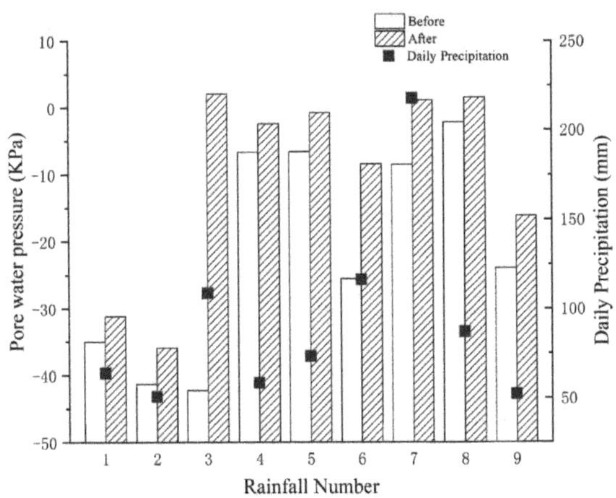

Figure 4. Change in pore water pressure subjected to heavy rainfalls at depth of 0.4 m in unscreened soil layer.

Different from the depths of 0.4 and 0.8 m, the increase in the pore water pressure at a depth of 1.2 m changed only slightly after each heavy rainfall. The wetter the CW layer, the larger the increase in the pore water pressure, for example, No. 3, 4, 5, and 8 rainfalls. The maximum increase in pore water pressure was 1 kPa. It may be because the wet CW layer increased the pore water pressure in the top boundary of the underlying screened soil layer.

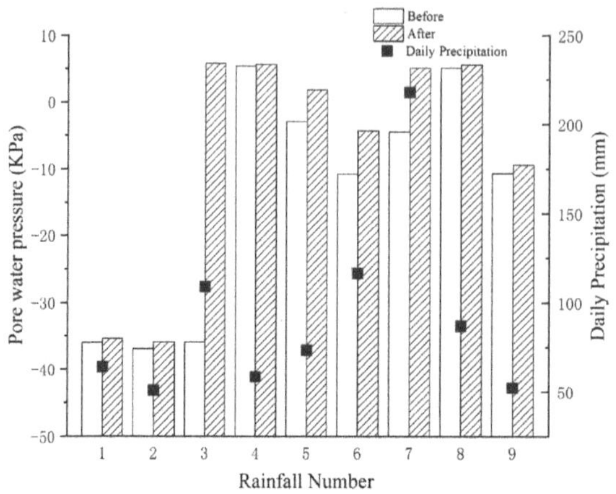

Figure 5. Change in pore water pressure subjected to heavy rainfalls at depth of 0.8 m in CW layer.

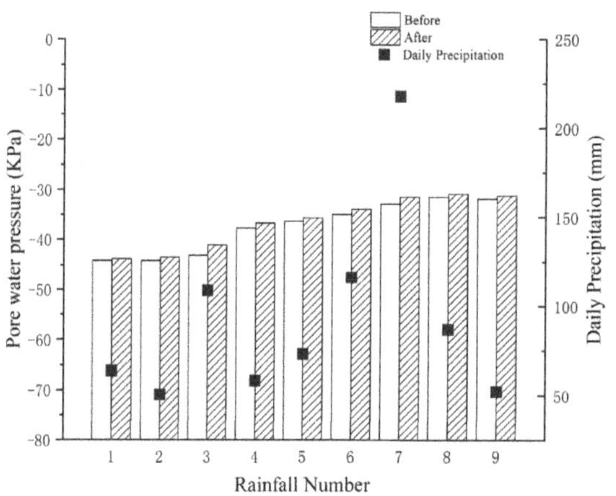

Figure 6. Change in pore water pressure subjected to heavy rainfalls at depth of 1.2 m in screened soil layer.

The No. 3, 7, and 8 rainfall events, although of different intensities (108.8 mm, 217.8 mm, and 86.9 mm), all caused the soil to reach the maximum pore water pressure, which will be maintained regardless of the amount of rainfall. Except for these three rainfalls, Figure 7 shows the relationship between daily precipitation and the increase in pore water pressure. As expected, increases in the daily precipitation and increases in pore water pressure in each soil layer were observed. The deeper the tensiometer, the smaller the ratio between the increase in pore water pressure and daily precipitation. As water seeped from the unscreened soil layer into the CW layer, this ratio decreased by approximately 48% (from 0.1771 to 0.0918). It was due to the capillary effect formed by the upper two layers. Once water seeped from the CW layer into the screened soil layer, the ratio decreased by approximately 98% (from 0.0918 to 0.0018). It was due to the low permeability of the screened soil. It indicates that the screened soil layer served as the major function for preventing water seepage in the three-layer cover system.

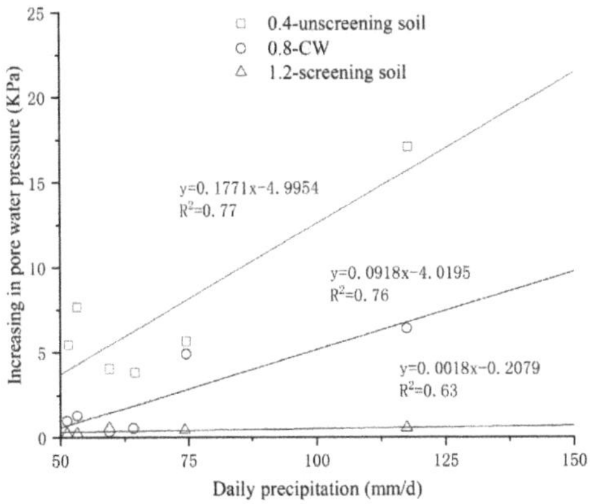

Figure 7. Relationship between daily precipitation and increase in pore water pressure except No. 3, 7, and 8 rains.

3.2. Percolation

Figure 8 shows the cumulative percolation at three sections. It should be noted that the cumulative percolation in a 6-month duration (i.e., from June to December) was about 10 mm. It indicates that the performance of the three-layer capillary barrier cover system using local soils and CW in the field condition was satisfactory under a humid climate where an annual rainfall is about 2000 mm.

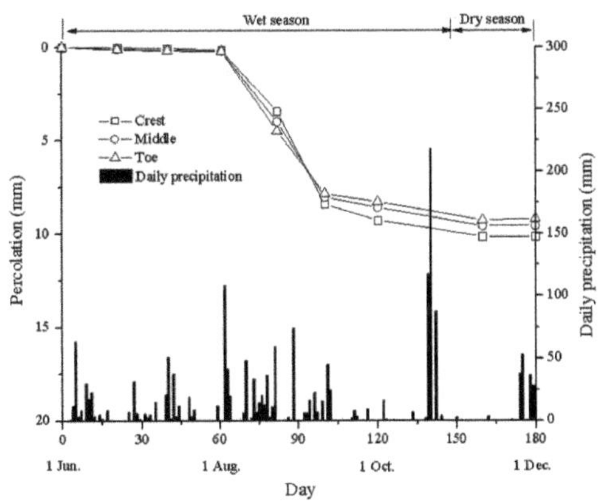

Figure 8. Development of percolation under daily precipitation condition.

From 1 June to 1 December 2016, the cumulative percolation development underwent distinct phases under varying rainfall conditions. In the initial phase (1 June–2 August), the initial pore water pressures from top to bottom were approximately −30 kPa, −35 kPa, −62 kPa, and −18 kPa. Characterized by prolonged weak rainfall, despite a total rainfall of 423.2 mm, the pore water pressure in the top two layers remained low, ranging from −25 kPa to −45 kPa at the 0.4 m (unscreened soil layer), while the pore water pressures in the other layers remained largely unchanged. No breakthroughs were observed, resulting in a negligible increase in cumulative percolation. The subsequent phase (2 August–5 September) experienced a combination of short-duration, high-intensity rainfall and continued prolonged weak rainfall. During this phase, the initial pore water pressures from top to bottom were approximately −30 kPa, −35 kPa, −60 kPa, and −18 kPa. After experiencing short-duration, high-intensity rainfall, the pore water pressures in the unscreened soil layer and CW layer increased to around 0 kPa, followed by a prolonged period of weak rainfall where the top two layers' pore water pressures remained persistently around zero. This condition led to breakthrough events, with a total rainfall of 602.2 mm and a significant cumulative percolation increase of approximately 8 mm. The final phase (5 September–1 December) was marked by a predominantly short-term intense rainfall, with initial pore water pressures from top to bottom of approximately −20 kPa, −10 kPa, −55 kPa, and −12 kPa. After experiencing a short-duration, high-intensity rainfall, the top two layers' pore water pressures briefly rose to around 0 kPa, with most pore water pressure readings remaining negative. Despite the total rainfall in this phase being higher at 687.8 mm than in the previous phase, the additional cumulative percolation was around 2 mm, which is less than the 8 mm increase observed in the second phase.

Regarding the percolation process, the pore water pressure in the CW layer, which serves as the top boundary of the screened soil layer, plays a crucial role. Figure 9 shows the relationship between percolation rate and average pore water pressure in the CW layer. As expected, the percolation rate increased as the average pore water pressure in the CW layer increased. It indicates that even in the period without rainfall, water will percolate when the CW layer is wet. According to the design criterion recommended by the US EPA, the percolation rate should not exceed 30 mm/year [30]. Following the U.S. EPA's ACAP guidelines, our study indicates that the CW layer's pore water pressure should remain below −3 kPa in humid climates.

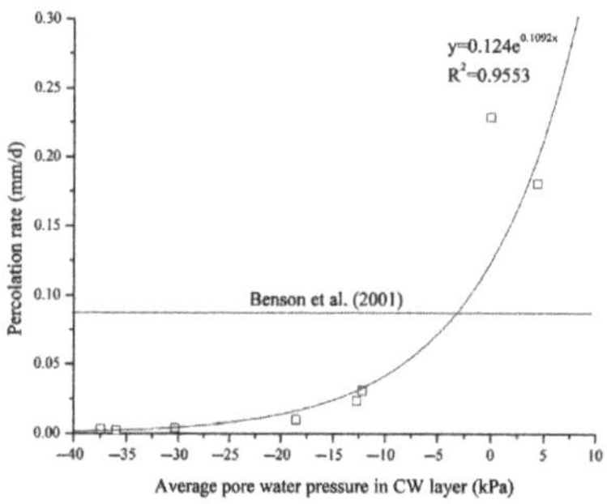

Figure 9. Relationship between percolation rate and average pore water pressure in CW layer [30].

4. Discussion

This field test examined a three-layer capillary barrier cover system, employing unscreened soil (25.2% fines content) and screened soil (40.0% fines content), combined with construction waste (CW), under natural rainfall. The focus was on the dynamics of pore water pressure and percolation across varying depths and under different rainfall intensities.

Below 1.2 m, pore water pressure remained largely constant, mirroring the simulated trends by Ng et al., (2015) [26], indicating minimal water infiltration to the lower layer. In high-rainfall scenarios, pore water pressures in the upper layers were positive, aligning with Ng et al.'s indoor model findings [2]. A notable deviation from these models was the slight rise in pore water pressure below 1.2 m during ongoing rainfall, attributed to increased pressure within the CW layer due to extended precipitation. Analysis revealed a direct proportionality between the percolation rate in the CW layer and the average pore water pressure.

Figures 4 and 8 reveal that prolonged weak rainfall, without breakthrough in the top two layers, results in minimal changes in cumulative percolation. However, breakthroughs occur when these layers transition from unsaturated to saturated states during short-duration, high-intensity rainfall, and continued weak rainfall prevents the layers from returning to an unsaturated state promptly, leading to percolation. Although short-duration, high-intensity rainfall was found to have a lesser impact on percolation, when followed by prolonged weak rainfall, it resulted in significant percolation. Despite the top two layers being in a saturated state, the CW layer, acting as a drainage layer, still effectively reduces percolation [4]. The effect of short-duration, high-intensity rainfall in elevating moisture levels substantially influences the landfill's response to further rainfall events.

The high relative humidity in landfill gas might influence the pore water pressure in the screened soil at the bottom [29], which could in turn affect landfill gas emissions [26].

The measured data and parameters of our study, conducted on a three-layer capillary barrier landfill cover system using local soils and construction waste, are subject to various environmental uncertainties. For example, measurement errors during heavy rain due to the inertia of tipping buckets, and factors like evaporation, wind effects, surface runoff generation, and lateral drainage were not fully incorporated in our boundary conditions. These elements could affect the system's performance under diverse environmental scenarios [31]. Future research should focus on quantifying these uncertainties to better predict system behavior.

Future research should focus on exploring the applicability of the cover system in different regions, particularly considering variations in soil and climate. Conducting long-term experiments under a variety of environmental conditions is crucial for understanding the system's response to climate change and aging. This includes a detailed analysis of the aging characteristics of materials, such as studying the changes in physical and chemical properties over time of materials used in the cover system, and how these changes affect the overall performance of the system. Additionally, the long-term monitoring of structural integrity is essential, requiring assessment of how structural damages (such as cracks and collapses) develop over time and impact the system's permeability and effectiveness.

5. Conclusions

A field test was carried out to study the performance of a three-layer cover system using local soils and construction waste. The detailed response of pore water pressure and percolation during rainfalls could be captured. Based on the measured results, the following conclusions may be drawn:

(1) In the wet season, the breakthrough of the upper two layers did not occur until a No. 3 rainfall whose daily precipitation was 108.8 mm. After the breakthrough, the upper two layers were wet until 15 September 2016. In this period, the pore water pressure in the underlying screened soil layer increased slightly and thus a major percolation occurred.

(2) The pore water pressure in the screened soil layer near the bottom was at a high level, and not correlated with daily precipitation. It may be due to the influence of the high humidity of the landfill gas.

(3) The increase in the pore water pressure in the upper two layers subjected to heavy rainfalls depended on not only the daily precipitation, but also the hydraulic state. Rainfall patterns significantly impact pore water pressure and percolation, with the combination of a short-duration, intense rainfall and prolonged weak rainfall notably extending the duration of saturated conditions and leading to significant percolation.

(4) Over the six-month period from June to December, the cumulative percolation in the study measured approximately 10 mm. These results indicate that the three-layer capillary barrier cover system, which integrates local soils and construction waste (CW), demonstrated initial effectiveness under the field conditions of a humid climate observed during the study period.

(5) The percolation rate increased as the average pore water pressure in the CW layer increased. In order to satisfy the allowable percolation rate of 30 mm/year, the pore water pressure in the CW layer should be smaller than −3 kPa.

Author Contributions: Conceptualization: J.L. and Y.W.; methodology: J.L. and Y.W.; software: J.R.; validation: J.R. and Y.W.; formal analysis: J.L., J.R. and Y.W.; investigation: J.L.; resources: Y.W. and J.R.; data curation: J.L. and J.R.; writing—original draft preparation: J.L. and J.R.; writing—review and editing: Y.W.; visualization: J.R.; supervision: Y.W.; project administration: J.L.; funding acquisition: J.L. All authors have read and agreed to the published version of the manuscript.

Funding: This work was supported in part by the National Natural Science Foundation of China under Grant 52378337.

Data Availability Statement: The data presented in this study are available in this article.

Conflicts of Interest: The authors declare no conflicts of interest.

Appendix A

Data Transmission System:

Figure A1 shows the architecture of the network system. The network system consisted of several analog-to-digital converters, a microcontroller, a data terminal unit (DTU) based on GPRS, a base transceiver station, a data center server, a computer, a solar panel, and a battery.

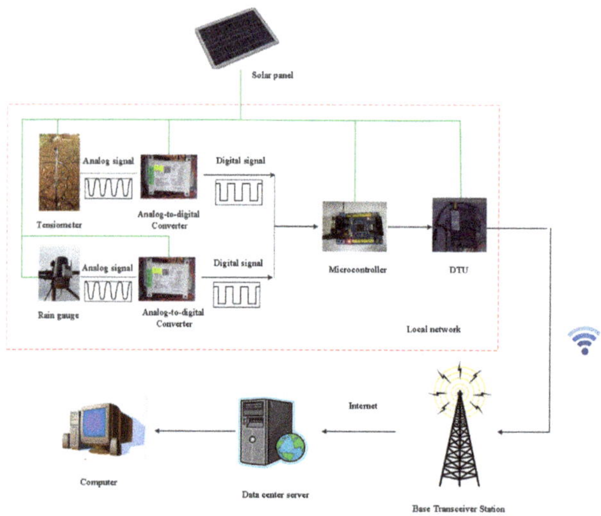

Figure A1. Architecture of network system.

In this architecture, the analog signal from each instrument was first sampled at regular time intervals in the analog-to-digital converter. Each sample was then evaluated in terms of a binary number and output in the form of the digital signal. This process may introduce errors, because a continuously varying analog signal was replaced by a discrete set of points at each time interval. In order to reduce the sample error to a level which could be accepted in field monitoring of the landfill cover system, the sample frequency was set as 1 Hz. The measuring range and accuracy of the analog-to-digital converter used in this study were ± 60 mV and 0.01 mV, respectively.

All the analog-to-digital converters were connected to a microcontroller to establish a local area network. In this network, each analog-to-digital converter was a slave whereas the microcontroller was a master. The microcontroller would send a request message to all the analog-to-digital converters and receive reply messages from the analog-to-digital converters with the desired address. All the received messages would be first saved to a storage unit in the microcontroller. It should be noted that even if the GPRS was not good, the local area network still worked. In other words, the data from each instrument were recorded and saved to the storage units and were independent of the GPRS network. On the other hand, the microcontroller was also connected to a DTU to form another area GPRS network. Based on this GPRS network, the microcontroller would send the data saved in the storage unit to a data center server. It should be noted that the results are saved in the storage unit during the period of poor GPRS quality.

The DTU is a wireless device which is applied to transmit the digital signal through the GPRS networks. Therefore, the DTU needed a build-in subscriber identification module (SIM) card which was used to apply for GPRS services. One of the advantages of GPRS networks is that it supports the DTU to permanently stay online. Another is that the GPRS network has covered almost all the provinces, municipalities, and autonomous regions. Once the DTU powered on, DTU first connected to a mobile station nearby and kept this communication link online. Subsequently, the digital signal from the analog-to-digital converter was read out by the DTU and saved in its cache. Finally, the DTU encapsulated the information in the cache into an information packet and sent the packet to a fixed internet protocol (IP) address which was configured in advance.

The base transceiver station is a piece of equipment that operates in the 3.5 GHz band for fourth-generation (4 G) mobile communication between user equipment (e.g., DTU, mobile phone) and a network. Via the communication, the base transceiver station could receive the information packet from DTU and then send it to the external public internet.

The data center server maintained online with a fixed IP address and could receive the information packets from DTUs. Once receiving the packets, the server would extract the useful data from the information packets and save it on the hard drive. When the data were required to be analyzed, a computer could be used to connect the server and download the data.

The solar panel and battery could supply enough electric energy for all instruments which were installed in the landfill cover. The solar panel first absorbed the sunlight as a source of energy to generate electricity. This electricity was then stored in a battery. The battery was connected to all instruments and supplies enough electricity for them.

References

1. Ng, C.W.W.; Liu, J.; Chen, R.; Lim, C.J. Numerical parametric study of an alternative three-layer capillary barrier cover system. *Environ. Earth Sci.* **2015**, *74*, 4419–4429. [CrossRef]
2. Ng, C.W.; Liu, J.; Chen, R.; Xu, J. Physical and numerical modeling of an inclined three-layer (silt/gravelly sand/clay) capillary barrier cover system under extreme rainfall. *Waste Manag.* **2015**, *38*, 210–221. [CrossRef] [PubMed]
3. Ng, C.W.; Coo, J.L.; Chen, Z.K.; Chen, R. Water Infiltration into a New Three-Layer Landfill Cover System. *J. Environ. Eng.* **2015**, *142*, 04016007. [CrossRef]
4. Ng, C.W.W.; Guo, H.; Ni, J.; Chen, R.; Xue, Q.; Zhang, Y.; Feng, Y.; Chen, Z.; Feng, S.; Zhang, Q. Long-term field performance of non-vegetated and vegetated three-layer landfill cover systems using construction waste without geomembrane. *Geotechnique* **2022**. *ahead of the print*. [CrossRef]
5. Albrecht, B.A.; Benson, C.H. Effect of Desiccation on Compacted Natural Clays. *J. Geotech. Geoenviron. Eng.* **2001**, *127*, 67–75. [CrossRef]
6. Barnswell, K.D.; Dwyer, D.F. Two-year performance by evapotranspiration covers for municipal solid waste landfills in northwest Ohio. *Waste Manag.* **2012**, *32*, 2336–2341. [CrossRef] [PubMed]
7. Zhan, G.; Lattin, D.; Keller, J.; Milczarek, M. 20 Years of Evapotranspiration Cover Performance of the Leach Pads at Richmond Hill Mine. *Mine Water Environ.* **2019**, *38*, 402–409. [CrossRef]
8. Rahardjo, H.; Santoso, V.A.; Leong, E.C.; Ng, Y.S.; Hua, C.J. Performance of an Instrumented Slope Covered by a Capillary Barrier System. *J. Geotech. Geoenviron. Eng.* **2012**, *138*, 481–490. [CrossRef]
9. Zhan, L.T.; Li, G.Y.; Jiao, W.G.; Lan, J.W.; Chen, Y.M.; Shi, W. Performance of a compacted loess/gravel cover as a capillary barrier and landfill gas emissions controller in Northwest China. *Sci. Total Environ.* **2020**, *718*, 137195. [CrossRef]
10. Shaikh, J.; Bordoloi, S.; Yamsani, S.K.; Sekharan, S.; Rakesh, R.R.; Sarmah, A.K. Long-term hydraulic performance of landfill cover system in extreme humid region: Field monitoring and numerical approach. *Sci. Total Environ.* **2019**, *688*, 409–423. [CrossRef]
11. Yu, Y.; Zhao, X.; Shi, Y.; Ou, J. Design of a real-time overload monitoring system for bridges and roads based on structural response. *Measurement* **2013**, *46*, 345–352. [CrossRef]
12. Kraimeche, B. Performance of a GSM/GPRS System Subject to Interference. In Proceedings of the 2005 5th International Conference on Information Communications & Signal Processing, Bangkok, Thailand, 6–9 December 2005; pp. 249–252. [CrossRef]
13. *ASTM D 422*; Standard Test Methods for Particle-Size Analysis of Soils. ASTM International: West Conshohocken, PA, USA, 2007.
14. *ASTM D 4318*; Standard Test Methods for Liquid Limit, Plastic Limit, and Plasticity Index of Soils. ASTM International: West Conshohocken, PA, USA, 2017.
15. *ASTM D 698*; Standard Test Methods for Laboratory Compaction Characteristics of Soil Using Standard Effort [12,400 ft-lbf/ft3 (600 kN-m/m^3)]. ASTM International: West Conshohocken, PA, USA, 2021.
16. *ASTM D 5084*; Standard Test Methods for Measurement of Hydraulic Conductivity of Saturated Porous Materials Using a Flexible Wall Permeameter. ASTM International: West Conshohocken, PA, USA, 2016.

17. Simion, I.M.; Fortuna, M.E.; Bonoli, A.; Gavrilescu, M. Comparing environmental impacts of natural inert and recycled construction and demolition waste processing using LCA. *J. Environ. Eng. Landsc. Manag.* **2013**, *21*, 273–287. [CrossRef]
18. Chen, Z.; Feng, Q.; Yue, R.; Chen, Z.; Moselhi, O.; Soliman, A.; Hammad, A.; An, C. Construction, renovation, and demolition waste in landfill: A review of waste characteristics, environmental impacts, and mitigation measures. *Environ. Sci. Pollut. Res.* **2022**, *29*, 46509–46526. [CrossRef] [PubMed]
19. Vasvári, V. Calibration of tipping bucket rain gauges in the Graz urban research area. *Atmos. Res.* **2005**, *77*, 18–28. [CrossRef]
20. Soilmoisture. *Model 2725A Jet Fill Tensiometer Operating Instructions;* Soilmoisture Equipment Corp.: Santa Barbara, CA, USA, 1997.
21. Chen, R.; Liu, J.; Ng, C.W.W.; Qiao, J. Development of a new assembled-type tensiometer. *Rock Soil Mech.* **2013**, *34*, 3028–3032. [CrossRef]
22. Feng, W.Q.; Liu, Z.Y.; Tam, H.Y.; Yin, J.H. The pore water pressure sensor based on Sagnac interferometer with polarization-maintaining photonic crystal fiber for the geotechnical engineering. *Measurement* **2016**, *90*, 208–214. [CrossRef]
23. Wang, L.; Liu, S.; Shen, C.; Lu, Y. Laboratory test and modelling of gas pressure under geomembrane subjected to the rise of groundwater in plain reservoirs. *Geotext. Geomembr.* **2020**, *49*, 81–96. [CrossRef]
24. Shenzhen Statistics Bureau, NBS Survey Office in Shenzhen. *Shenzhen Statistical Yearbook;* China Statistics Press: Beijing, China, 2017.
25. Wu, M.; Lam, H.; Li, K. Characterization and indexing of heavy rainstorms in Hong Kong. *Meteorol. Appl.* **2015**, *22*, 25–36. [CrossRef]
26. Ng, C.W.; Liu, J.; Chen, R. Numerical Investigation on Gas Emission from Three Landfill Soil Covers under Dry Weather Conditions. *Vadose Zone J.* **2015**, *14*, vzj2014.12.0180. [CrossRef]
27. Avzalshoev, Z.; Uchimura, T. An experimental approach to investigating Quasi-Saturation using Darcy's Law. *Sustainability* **2023**, *15*, 10459. [CrossRef]
28. Yang, J.; Ma, X.Q. The benefit of landfill gas power generation to the reduction of the greenhouse gas emission. *Environ. Pollut. Control* **2006**, *28*, 461–464.
29. Piechota, G.; Hagmann, M.; Buczkowski, R. Removal and determination of trimethylsilanol from the landfill gas. *Bioresour. Technol.* **2012**, *103*, 16–20. [CrossRef] [PubMed]
30. Benson, C.; Abichou, T.; Albright, W.; Gee, G.; Roesler, A. Field Evaluation of Alternative Earthen Final Covers. *Int. J. Phytoremediat.* **2001**, *3*, 105–127. [CrossRef]
31. Baker, E.A.; Manenti, S.; Reali, A.; Sangalli, G.; Tamellini, L.; Todeschini, S. Combining noisy well data and expert knowledge in a Bayesian calibration of a flow model under uncertainties: An application to solute transport in the Ticino basin. *arXiv* **2022**, arXiv:2210.17388. [CrossRef]

buildings

MDPI

Article

Evaluation Model of Distributed Photovoltaic Utilization in Urban Built-Up Area

Siyuan Chen [1,2], Zao Zhang [1,*], Cheng Wang [3,*], Lifeng Tan [4], Huanjie Liu [1], Hong Yuan [1], Rui Zhang [1] and Rui Hu [1]

1 School of Architecture, Tianjin University, Tianjin 300384, China; liuhuanjie_arch@163.com (H.L.)
2 China Northeast Architectural Design & Research Institude Co., Ltd., Shenzhen 518057, China
3 School of Art, Tianjin Chengjian University, Tianjin 300384, China
4 School of Architecture, Tianjin Chengjian University, Tianjin 300072, China; tanlf_arch@163.com
* Correspondence: zao.zhang@tju.edu.cn (Z.Z.); cheng_gsa@163.com (C.W.)

Abstract: Photovoltaic (PV) power generation is emerging as a key aspect of the global shift towards a more sustainable energy mix. Nevertheless, existing assessment models predominantly concentrate on predicting the overall capacity of PV power generation, often neglecting temporal dynamics. Drawing upon the urban energy substitution rate, utilization rate, and power supply stability, this study has devised a comprehensive evaluation model for the utilization of distributed photovoltaic systems (SUS). This model integrates the quantification of spatio-temporal features inherent in urban settings and buildings. Using Hohhot city as a case study, this study conducted simulations to analyze how the installation of PV systems affects the electricity consumption patterns across different land plots within the urban core. The study additionally examines how urban planning influences the adoption of PV power, taking into account both the timing of PV power usage and the stage of PV technology development. The evaluation model surpasses the constraints of current urban PV assessments, which primarily emphasize enhancing power generation potential without adequately quantifying supply–demand dynamics or spatial and temporal variations. This breakthrough significantly improves the precision and reliability of assessing the efficiency of distributed PV systems. Its implications extend widely to subsequent comprehensive evaluations of urban PV applications.

Keywords: distributed photovoltaic; evaluation model; sustainable energy; quantitative analysis

1. Introduction

Today, 80% of the world's GDP comes from 2% of the world's cities, but they consume nearly 75% of the world's resources [1]. Renewable energy has become a crucial issue that must be addressed in the national development process, as energy security is directly linked to economic growth [2,3].

As of 2018, the total global installed PV capacity has grown to 480 GW (from only 1 MW in 1979), accounting for approximately 2% of global energy consumption [4]. This demonstrates a rapid development trend. In the future, until 2050, PV power generation will continue to drive the structural transformation of renewable energy worldwide due to its abundant resource availability, substantial market potential, and cost competitiveness [5]. One of the most important forms of photovoltaic power generation is distributed photovoltaic power generation, which involves installing photovoltaic modules on the surfaces of buildings in cities to generate electricity [6].

However, regarding distributed photovoltaics in urban built-up areas, energy spatial planning and urban planning will both play crucial limiting roles [7]. The current indicators focus on enhancing the production potential of generation, but there are significant shortcomings in considering the energy supply-and-demand relationship, as well as spatial and temporal variations.

In the past, the assessment of urban PV potential focused on the installable area for PV systems and the power generation capacity in different urban areas. Firstly, evaluation indicators should be established. At present, several common evaluation indicators or standards are shown in Table 1. However, the established research indicators are relatively single, only consider the total amount of power generation/total amount of power consumption, ignoring the temporal sequence of PV power consumption and the stage of development.

Table 1. Main application indicators of current urban PV potential assessment work.

Indicators	Definition of Indicator	Index Source
PV area per capita	Area PV available roof area/area total population	IEA-PVPS [8]
Per capita installed PV capacity	Total installed PV capacity/total population of the region	Izquierdo, S. [9]
$SAFAR_n$	Photovoltaic available area on building exterior surface/building area	Jouri, K. [10]
Photovoltaic permeability	Regional total installed PV capacity/regional peak load	Zhao, B., Xiao, C.H., et al. [11]
PV cost/PV utilization	Regional photovoltaic installation cost and electricity sales revenue by weight comprehensive comparison	Cheng, M., Zhao, S.H., et al. [12]
Solar energy grading factor	Area available for photovoltaic power generation/area of land use	Sigrid Lindner of Ecofys [13]
RPVID	Area of photovoltaic available on the regional roof/area of land use	Zhang, H. [14]

In addition to the relevant indicators, the methodology for potential assessment is at the core of the research. The urban PV potential is usually assessed by giving a percentage of the building surface that is suitable for urban PV development (usually the roof utilization factor). In terms of the scope of the assessment, calculations are conducted for specific building types. For instance, Enongene et al. analyzed and modeled typical buildings to calculate the PV potential of urban areas in Nigeria [15]. There are also projections on a national scale. For example, Rosas-Flores et al. assessed the energy potential of Mexican city pools and rural households through large-scale grid-connected and interconnected PV systems [16]. Bodis et al. assessed the potential PV power generation for the entire European Union, estimating 680 TWh per year [17]. In terms of assessment calculation methods, accurate quantitative studies incorporating GIS systems have also been used by an increasing number of scholars for assessing urban PV potential [18–21]. Alaa Alhamwi et al. proposed the establishment of a GIS open-source platform for optimizing urban distributed energy systems [22]. Mariela Tapia et al. utilized GIS and associated building feature extraction algorithms to evaluate rooftop solar PV potential in various cities in Ecuador [23]. Jiang Liu et al. combined 3D building data with GIS to accurately assess the solar PV potential of rural roofs and facades [24]. Most of these potential assessment studies focus on a city scale or specific building types to maintain the high accuracy provided by the aforementioned methods. There are also uncertain predictions for intricate cities, such as the study by Diana Bernasconi and Giorgio Guariso on surface quantities and the potential of rooftop PV in complex cities. This study utilized morphological clustering methods instead of relying on remote sensing data for large-scale studies [25].

Most of the previous studies on PV potential have focused on estimating the capacity of PV power generation by predicting the ratio of PV power production to the current urban electricity consumption. However, the discussion focusing on the theoretical increase in the total power generated has also contributed to overly optimistic expectations. On the one hand, PV power output is greatly influenced by environmental factors. Without considering the impact of specific climatic conditions such as clouds and rain, PV power output typically peaks at midday and follows a clear normal distribution. As the primary source of urban energy consumption, the pattern of electricity load change is associated with users' behavioral characteristics and exhibits variations based on the diverse functions of buildings. The two do not always show a similar trend of change due to the simultaneity of power output and consumption behavior. The excess PV power output cannot be effectively consumed immediately.

The primary solution to this problem is to depend on different energy storage technologies to maintain the balance between system supply and demand, thereby enhancing its overall stability. However, due to cost constraints, energy storage facilities cannot be widely used for commercial purposes in urban areas. This creates a significant technological bottleneck that hinders the utilization of photovoltaics [26]. The excess PV power can be transferred to other urban areas through grid-connected methods. However, in the future, when the city relies on PV energy as the primary source of electricity, the excess power cannot be transferred and consumed within the city. Therefore, it is necessary to propose a dynamic indicator to introduce the concept of PV consumption timing. This indicator, known as the PV Utilization Assessment Method (SUS), is based on the energy substitution rate, PV utilization rate, and PV utilization stability. It aims to comprehensively reflect the current status of PV application and integration in all regions of the city.

2. Materials and Methods

2.1. Research Materials

In the research of distributed PV in urban built-up areas, we selected the central city of Hohhot as the primary research object (Figure 1). Hohhot is classified as a second-tier region for solar energy sunshine resources in China, indicating better potential for PV development. It is considered a priority region for large-scale PV installations in the future. At the same time, Hohhot's three types of industrial structure ratios are very similar to those of China, and its urban function ratio can reflect the characteristics of China's urban economic development.

Figure 1. Location of the study area. (The map of China (left) was provided by Ministry of Natural Resources of China (Approval number: GS(2019)1673)).

Hohhot city in Inner Mongolia is located between 110°46′–112°10′ E and 40°51′–41°8′ N. It belongs to the primary solar resource distribution area of China. The total annual solar radiation is 6241.19 MJ/m^2 (about 1734 kWh/m^2), and the effective sunshine time is 2599.7 h. This indicates promising prospects for PV development. Meanwhile, in terms of electricity consumption load in Hohhot, the total area of various types of construction land reached 329,554,000 square meters. The entire society in Hohhot used 33,084,270,000 kWh of electricity in 2022. The city mentioned above has a high electricity demand, coupled with abundant sunshine energy, that holds great potential for development [27].

The information on the urban built-up area of Hohhot was obtained from the planning data and relevant policies and regulations of the relevant departments. The building load information was calculated and summarized based on the meter readings provided by the state power department. In addition, all relevant data, including the comparison of photovoltaic technology characteristics, the land-cover status of centralized photovoltaic land, the data on protected areas and global sunshine radiation, the distribution of population density, per capita electricity consumption, meteorological data, and other information, were sourced from open data provided by countries and various global organizations.

2.2. Research Methods

2.2.1. Formulate SUS PV Potential Assessment Indicators

Energy Substitution Rate

The concept of energy substitution rate (ESR) considers the time sequence and actual situation of photovoltaic electricity consumption to reflect the degree of substitution of renewable energy for traditional electricity consumption. It measures the degree of spontaneous self-use after the integrated application of photovoltaic systems in various urban areas. Specifically, it can be divided into the overall energy substitution rate (ESR) and the hourly energy substitution rate (ESR_i) of regional plots, as shown in Equations (1) and (2).

The hourly energy substitution rate (ESR_i) is as follows:

$$ESR_i = \frac{Pe_i}{Load_i} \tag{1}$$

The overall energy substitution rate (ESR) is as follows:

$$ESR = \frac{\sum_i^n Pe_i}{\sum_i^n Load_i} \tag{2}$$

ESR_i is the hourly energy substitution rate of a block area in the city at the ith time. ESR is the overall energy substitution rate in the period from 0 to 23 points in a certain area of the city. Pe_i (kWh) is the consumption of photovoltaic power produced by a block in a city at the ith time. $Load_i$ (kWh) is the electricity consumption load at the ith moment of a block in a city. PV_i (kWh) is the photovoltaic power generation at the ith moment of a block in a city.

Photovoltaic Utilization Rate (PUR)

In addition to considering the replacement degree of traditional energy sources, the daily load curve patterns in various urban areas will change according to the different building functions (residential, office, commercial, etc.). In this paper, the photovoltaic utilization rate [28] (PUR, PV utilization ratio) is a supplement to the PV utilization ratio index (ESR), which is defined as the ratio of actual available photovoltaic power generation to the maximum allowable power generation under theoretical conditions within a specific operating period.

$$PUR = \frac{\sum_i^n Pe_i}{C \cdot T} \tag{3}$$

PUR is the photovoltaic utilization rate of a certain area in the city. C (kWh) is the installed photovoltaic capacity of a certain area in the city. T (h) is the theoretical working hours of PV module in a certain area of the city.

Standard Deviation of Hourly Replacement Rate

Photovoltaic power generation technology is significantly influenced by climate change factors. The stability of power supply output exhibits considerable volatility with changes in weather and time. This instability is evident in the variation of the photovoltaic daily hourly replacement rate (ESR_i) over time. In this study, the standard deviation (Stdev) of the hourly PV replacement rate was used to reflect the power supply stability of each urban parcel after PV application, as shown in Formula (4):

$$\text{Stdev} = \sqrt{\frac{1}{n} \cdot \sum_{i}^{n} (\text{ESR}_i - \text{R}_i)^2} \tag{4}$$

Stdev is the standard deviation of the hourly PV replacement rate. R_i is the arithmetic average of daily hourly PV replacement rate. n is the number of power generation hours during the photovoltaic available period of the plot.

2.2.2. SUS PV Potential Assessment Model

The PV potential assessment of SUS mainly includes four steps: basic data collection, data processing, SUS index, and formulation of regional planning and adjustment strategies.

The basic data collection part includes load information data such as, daily load rate curves of various buildings and their peak load densities; urban model information data such as, the building floor area ratio, building density, and functional ratio of each building; and meteorological information data such as, annual sunshine hours and annual hourly sunshine radiation intensity.

Firstly, the corresponding hourly load and hourly power generation of each block are calculated. Based on this data, the ESR, PUR, and Stdev indexes of each block are calculated, respectively.

Finally, through the classification and analysis of the three indicators, the priority of photovoltaic development in each region is evaluated. In this way, it can be determined which land parcel's functional form is more suitable for the future energy structure transformation. Then, the location layout and urban planning can be adjusted for the unfavorable areas of photovoltaic development to optimize the overall utilization of photovoltaics in the city (Figure 2).

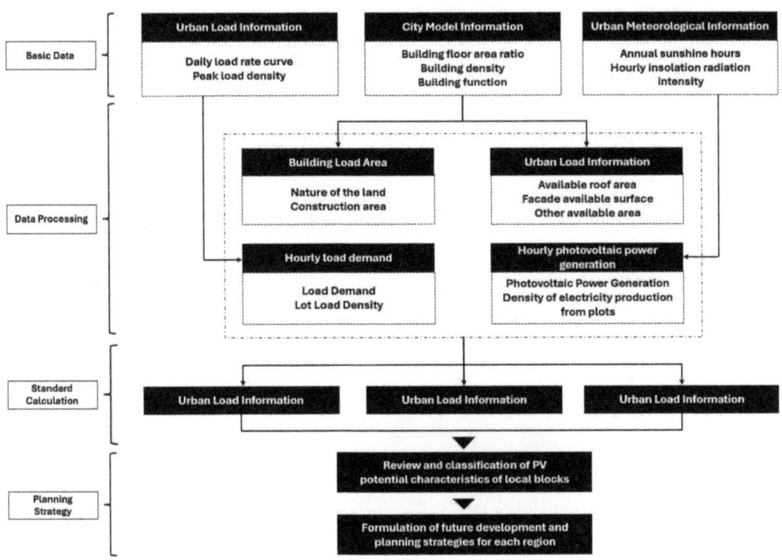

Figure 2. Framework diagram of SUS PV potential assessment procedure. Source: Drawn by the author.

2.2.3. Urban Basic Form Calculation

Based on the overall urban plan and zoning regulations of Hohhot City for 2020–2030, this paper examines the three indices of building function, building density, and floor area ratio for each block within the designated planning area.

Building Volume Rate for a Certain Area

The basic data was provided by the Hohhot Urban and Rural Planning and Design Institute. The overall plan (2020–2030) includes vector information such as land use boundaries, land use characteristics, and road network planning for various areas. The zoning plan includes Huhhot Hui District, Saihan District, Xincheng District, and Yuquan District. The controlled range of land use intensity (plot ratio) is as follows: A plot ratio >4.5, B plot ratio 3.5–4.5, C plot ratio 2.0–3.4, and D plot ratio <2.0 (Figure 3).

Figure 3. Plot ratio simulation results of each block. Source: Drawn by the author. Building density of a certain area.

Building density (BD) refers to the ratio between the total base area of a building and the land area of a plot within a certain range. It is an important indicator that reflects the building form information of different blocks. Since it is still in the planning stage, this paper adopts a calculation method for building density based on the characteristics of a typical urban plot layout. The building density of each block can be roughly calculated by Equation (5):

$$BD' = (1 - G - R - S - P) \times 100\% \tag{5}$$

BD' means the preliminary estimated building density of a certain block. G is the green land rate of a place. R is the land use rate of a block of roads. S is the rate of supporting public construction land for a residential parcel. P is the percentage of ground parking land in a certain area.

Draw up the Number of Building Floors and Adjust Building Density

According to Equations (4) and (5), the building density (BD') of each block can be preliminarily estimated. At the same time, according to the relationship between floor area ratio and building density, the average floor number (F′) of each block building can be further calculated. To facilitate further prediction and analysis, we assume that buildings in each given land parcel have the same number of building floors and in turn adjust the preliminary calculated building density (BD′) results for each parcel. Finally, the proposed building density (BD) and the number of building floors (F) of each block can be used for calculation. The calculation process is shown in Equations (6)–(8).

$$f' = \frac{FAR}{BD'} \quad (6)$$

$$f = roundup(f' + 0.5) \quad (7)$$

$$BD = \frac{FAR}{f} \quad (8)$$

FAR is the volume ratio of a certain area. F′ means the average number of building floors preliminarily calculated for a certain block. BD′ is the building density preliminarily calculated for a certain block. f is the proposed number of building floors after adjustment of a certain block and is the result of f′ rounded up. BD is the proposed building density after the adjustment of a place block.

2.2.4. Calculation of Available PV Area

Available Rooftop PV Area

The rooftops of various buildings in the city are the primary locations for photovoltaic applications. For specific buildings, the available roof area can be determined through software simulation or manual screening. However, for a large range of urban areas, it is usually calculated by the rooftop photovoltaic utilization coefficient (Kr). The roof utilization coefficient refers to the research results of Zhang Hua in 2017 (Table 2). The available area of rooftop PV (Sr) of each building block can be calculated by Equation (9).

$$Sr = BD \times A \times Kr \quad (9)$$

Sr (m^2) is the available rooftop photovoltaic area of a block building. A (m^2) is the land area of a certain place. Kr is the rooftop photovoltaic utilization coefficient corresponding to a block type.

Table 2. PV roof utilization factor at a glance and the values taken in this paper [29].

Land Use Category		City Discount Rate	Neighborhood Discount Rate	Monomeric Discount Rate	Rooftop PV Utilization Factor (Kr)	Mean Value (Kr) Values
Residential buildings	Low-rise residential area	1	1	0.3–0.5	0.3–0.5	0.4
	Multi-story residential area	0.85–1	0.95–0.98	0.55–0.7	0.44–0.69	0.57
	High-rise residential area	0.9–1	0.7–0.95	0.4–0.6	0.25–0.57	0.41
Multi-story public buildings		1	1	0.7–0.85	0.7–0.85	0.78
High-rise public buildings		0.85–1	0.51–0.8	0.5–0.7	0.22–0.56	0.39
Single/low-rise industrial buildings		1	1	0.8–0.9	0.8–0.9	0.85

Assessment of Photos

According to the layout characteristics of general urban buildings, each block area can be divided into two layout modes, multi-block layout and single-family/tower layout, to simplify the complex changes in urban building forms, as shown in Figure 4. Among them, the former is applicable to the general land use properties such as low multi-story/small high-rise buildings with an average building level (F) below 10, while the latter is applicable to land use properties of individual buildings such as commercial buildings, wholesale markets, and high-rise tower layouts with an average building level (F) above 10.

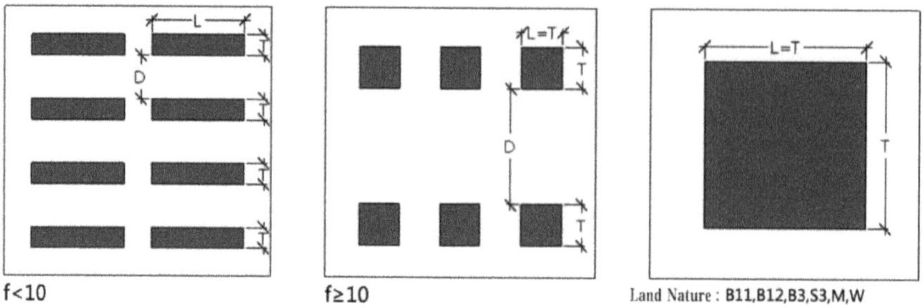

Figure 4. Simplified urban plot layout: Multi-block layout (**left**), tower layout (**middle**), single-family layout (**right**). Source: Drawn by the author.

The azimuth angle of the layout of each block is oriented towards the south to ensure that the south facade can receive maximum solar radiation. When each block has the same land use property, if multiple buildings are arranged in a row, each building has the same building depth (T). For energy conservation, single-family/tower layout buildings have the same building depth (T) and face width (L) to minimize their building size factor [30]. The values of general depth (T) and height (H) of reference buildings are determined from relevant codes, design regulations, and design experience values (Table 3). The area of the south facade of each block in the city can be roughly and statistically evaluated using Equations (10) and (11).

Table 3. In this paper, the depth (T) and height (h) of each type of building are taken from [31–33].

Nature of Land Use	Site Code	Floor Height h (m) Average	Depth of Entry T (m)
Administrative office land	A1	3.9	16
Cultural facilities land	A2	4.2	16
Educational and research land	A3	3.9	18
Primary school land	A33	3	10
Medical and health land	A5	3	24
Social welfare land	A6	3.6	16
Retail commercial land	B11	4.5	Single building layout T = L
Mixed commercial/hotel land	B11/B14	3.9	20
Mixed commercial/office land	B11/B2	3.9	16
Wholesale market land	B12	4.5	Single building layout T = L
Hotel land	B14	3.9	20
Business office land	B2	3.9	16
Recreation and sports land	B3	4.5	Single building layout T = L
Public utility business land	B4	3.9	18
Other service facilities land	B9	3.9	18
Industrial land	M1	6	Single building layout T = L
Residential land	R	3	15
Residential/commercial land	R/B11	3.2	15
Comprehensive transportation hub land	S3	4.5	Single building layout T = L
Storage and logistics land	W	6	Single building layout T = L

Statistics for the area of the south facade of multiple townhouses are calculated as follows:

$$\mathrm{Sn} = \mathrm{f} \cdot \mathrm{h} \cdot \frac{\mathrm{Ar}}{\mathrm{T}} \tag{10}$$

Statistics for the area of the south facade of a single-family/tower layout are calculated as follows:

$$\mathrm{Sn} = \mathrm{f} \cdot \mathrm{h} \cdot \sqrt{\mathrm{Ar}} \tag{11}$$

Sn (m^2) is the statistical value of the south elevation area of a block. f is the number of proposed building floors for a certain block. h (m) is the general building height selected according to the nature of land. Ar (m^2) is the roof area/floor area. T (m) is the general building depth selected according to the nature of land use.

Describe the window reduction coefficient (Kf) of the facade and the shadow reduction coefficient (Ks) of the facade. That is, the photovoltaic available area (Sf) of the south facade of each block can be expressed as follows:

$$\mathrm{Sf} = \mathrm{Sn} \cdot \mathrm{Kf} \cdot \mathrm{Ks} \tag{12}$$

For the value of Kf, we ignored the influence of shadow occlusion caused by the changes in the concave and convex shape of building facades, such as shading and projection, and calculated it only with the area ratio of window to wall (100%—window–wall area ratio). In combination with relevant codes [34], the value of Kf can be set as 0.55 for general buildings and as 1 for commercial buildings (B11). However, for general commercial building plots (B11) or mixed plots including commercial functions (B11/B14, B11/B2, R/B11), the first 1–2 floors of commercial building parts are usually not suitable for photovoltaic module installation due to their extremely high commercial value and should be excluded in the assessment of photovoltaic available facade area. In other words, the value of the proposed building floor (F) in each commercial and mixed parcel containing commerce in Equations (10) and (11) shall be considered by subtracting 1–2 floors.

As for the value of Ks, according to the standard, the sunshine duration of 6 h during the winter solstice should be guaranteed [35]. Therefore, in this paper, the elevation shadow reduction coefficient (Ks) of the multi-block layout was calculated at the most unfavorable time point (9:00 a.m. on the winter solstice). As shown in Figure 5, the ratio of photovoltaic available facade area to facade area (Ks) can be expressed as AB/AC and can also be expressed as C′c/C′o, according to the similar triangle theorem. The detailed calculation process is shown in Equations (13)–(16).

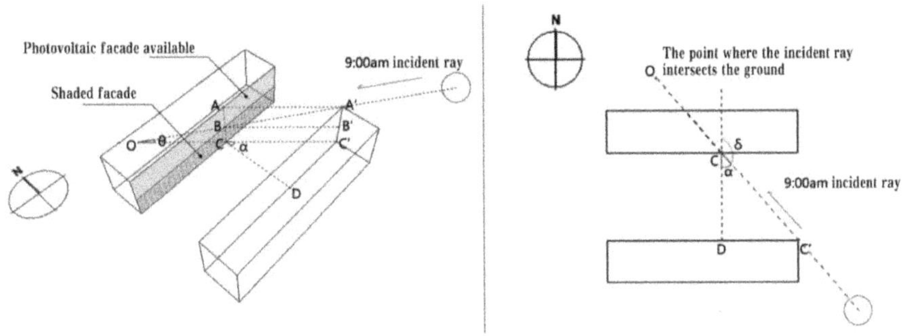

Figure 5. Calculation of the shadow reduction coefficient Ks of the multi-block layout facade. Source: Drawn by the author.

The Ks calculation process of multi-building row layout is determined as follows [35]:

$$\mathrm{Ks} = \frac{\mathrm{AB}}{\mathrm{AC}} = \frac{\mathrm{A'B'}}{\mathrm{A'C'}} = \frac{\mathrm{B'B}}{\mathrm{C'O}} = \frac{\mathrm{C'C}}{\mathrm{C'O}} \tag{13}$$

$$\mathrm{C'C} = \frac{\mathrm{CD}}{\cos(180^\circ - \delta)} = \frac{\mathrm{L0} \cdot \mathrm{h} \cdot \mathrm{f}}{\cos(180^\circ - \delta)} \tag{14}$$

$$\mathrm{C'O} = \frac{\mathrm{A'C'}}{\tan\theta} = \frac{\mathrm{h} \cdot \mathrm{f}}{\tan\theta} \tag{15}$$

By substituting Equations (14) and (15) into Equation (13), we can get the following:

$$\text{Ks} = \frac{\text{L0}\cdot\tan\theta}{\cos(180^\circ - \delta)} \tag{16}$$

Ks is the shadow reduction coefficient of the south-central facade of a multi-block layout. L0 is the reference sunshine spacing coefficient. h (m) is the general building height selected according to the nature of the land. f is the number of proposed building floors. δ means the azimuth angle of the sun at a certain time. θ is the height angle of the sun at a certain time.

At the same time, it is known that the sunshine spacing coefficient (L0) of Hohhot City is 1.73 (2 h on a major cold day) [36], the sun azimuth angle (δ) is 138.19° at 9:00 a.m. on the winter solstice, and the sun altitude angle (θ) is 13.34° (calculated according to the method in the literature [37]). By substituting Equations (5)–(15), we can calculate that the value of the facade shadow reduction coefficient (Ks) for the multi-building row layout in Hohhot is 0.55.

Assessment of Photovoltaic Available Area of a Parking Lot Surface

Approximately 10% of the land area of the parking lot and 20% of the canopy area cannot be utilized for photovoltaic installation. Finally, the available photovoltaic area of various parking lots can be expressed by Equation (17). The calculation results are shown in Figure 6.

$$\text{St} = 0.72\cdot\text{A}\cdot\text{P} \tag{17}$$

St (m^2) is photovoltaic available area of a parking lot. P is the rate of ground parking land, assuming that the social parking lot and bus station are 100%.

Figure 6. Total available PV area of various blocks in Hohhot (roof, south facade, and ground parking area). Source: Drawn by the author.

2.2.5. Evaluation of Hourly Photovoltaic Power Generation

Under the condition that the available PV area of each block in the city is known, we can evaluate and calculate the hourly electricity production of each area. Considering the feasibility of urban distributed photovoltaic installations in the future, this paper selects a relatively common and cost-effective polysilicon solar cell module (with an average photoelectric conversion efficiency of about 16%) for calculation. It is also assumed that the roofs, facades, and ground parking areas of block buildings around the city are all installed with a relatively common tilt, and photovoltaic applications are carried out in a south-facing arrangement. The optimal tilt angle (β) of a photovoltaic installation is 37.78° (latitude 40.78° for Hohhot) according to the relevant provisions of the grid connection system. This is calculated by subtracting 3 from the local latitude, as specified in the design specification of a photovoltaic power station.

In terms of statistical capacity per unit area (Cp) [38], we consulted the relevant specification for horizontal photovoltaic installations (roofs, various ground parking areas) and photovoltaic facade installations. We selected two distinct points in time during the year that represent the most unfavorable conditions to assess the spacing of photovoltaic modules, photovoltaic coverage, and capacity per unit area for each installation type.

The statistics of installed capacity per unit area can be calculated by determining the ratio between the projected area of PV modules and the available area of PV, that is, the PV coverage rate (Pr). This paper briefly describes the indicator as the ratio between the PV module projection length and the PV array spacing, as illustrated in Figures 7 and 8. The PV coverage of the horizontal plane and elevation can be calculated by referring to Equations (13)–(16). After substituting the local parameters in Hohhot, (at 9:00 on the winter solstice, the sun's altitude angle $\theta = 13.34°$ and the sun's azimuth angle $\delta = 138.19°$; at 12:00 on the summer solstice, the solar altitude angle $\theta = 72.62°$, and the solar azimuth angle $\delta = 180°$), it can be determined that the photovoltaic coverage rate of the horizontal plane is 75.4% and the photovoltaic coverage rate of the facade is 19.5%. According to Equations (18) and (19), the installed capacity per unit area of the horizontal plane (Cph) in Hohhot is 158 W/m^2, and the installed capacity per unit area of the facade (Cpv) is 53 W/m^2.

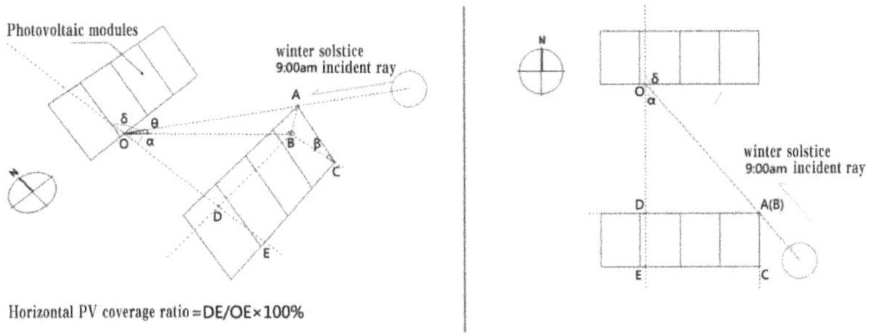

Figure 7. Calculation of the PV coverage when installed on an inclined horizontal plane. Source: Drawn by the author.

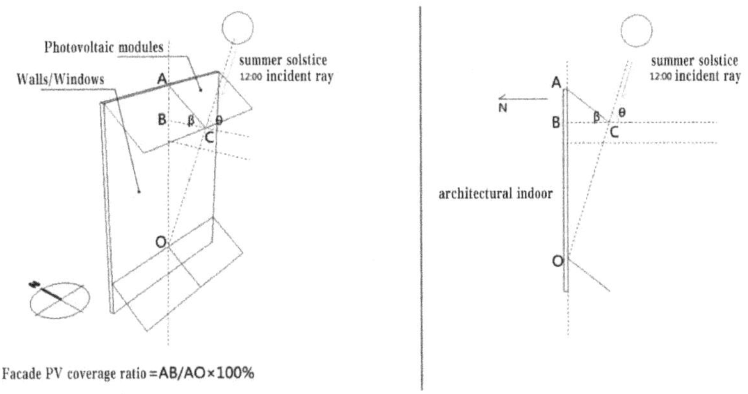

Figure 8. Calculation of the PV coverage when installed on a tilted facade. Source: Drawn by the author.

The installed capacity per unit area of the horizontal plane is as follows [35]:

$$\text{Cph} = \frac{\text{Pw}\cdot\text{Pr}}{\text{Ap}\cdot\cos\beta} \tag{18}$$

The installed capacity per unit area of facade plane is determined by the following:

$$\text{Cpv} = \frac{\text{Pw}\cdot\text{Pr}}{\text{Ap}\cdot\sin\beta} \tag{19}$$

Cph (W/m^2) is the estimated installed capacity per unit area of the horizontal plane of the proposed area. Cpv (W/m^2) is the estimated installed capacity per unit area of the proposed area facade. Pw (Wp) means the peak power (rated power) of the selected PV module. Pr is the PV coverage rate of the PV module installation area. Ap (m^2) is the area of the selected PV module. $\boldsymbol{\beta}$ is the best angle selected for photovoltaic module installation (37.78° in Hohhot).

3. Results and Discussion

3.1. Hourly Electricity Generation Calculation

After determining the installed capacity per unit area of the horizontal plane and facade, we can proceed to calculate the hourly power generation per unit area of photovoltaic panels. Firstly, the hourly meteorological forecast data for Hohhot City from 0:00 on 1 January 2019 to 23:00 on 31 December 2019 was selected from EnergyPlus. Then, we combined the SAM (System Advisor Model) PV Model analysis software (2015.1.30) provided by the United States National Renewable Energy Laboratory (NREL) to analyze the variation in hourly PV output power per unit area on the horizontal plane and facade in Hohhot city throughout the year (Figure 9).

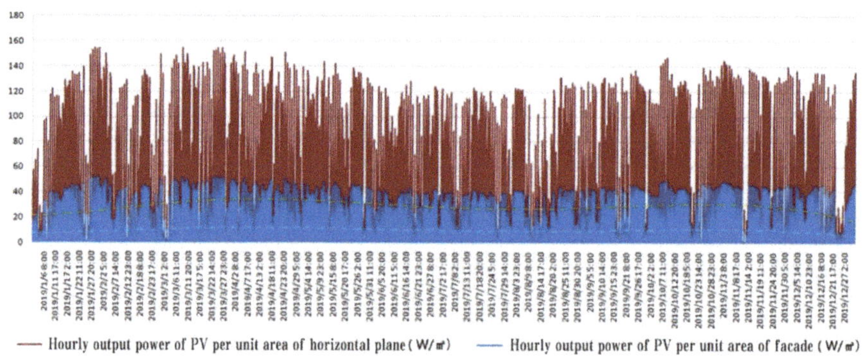

Figure 9. Annual hourly PV output per unit area of horizontal plane and facade of Hohhot. Source: exported by the author according to the analysis result of SAM.

The results indicate that the overall variation throughout the year remains stable, with a slight increase in photovoltaic electricity production observed during the periods from March to May and September to November. By utilizing SPSS software (26.0), we conducted k-means cluster analysis on the hourly PV output per unit area of Hohhot City throughout the year on a daily basis. This analysis helped us categorize the days with similar characteristics in terms of hourly PV output fluctuations. With this clustering (K = 10), we obtained a 10-category clustering result (as shown in Figure 10, arranged in descending order of occurrence days).

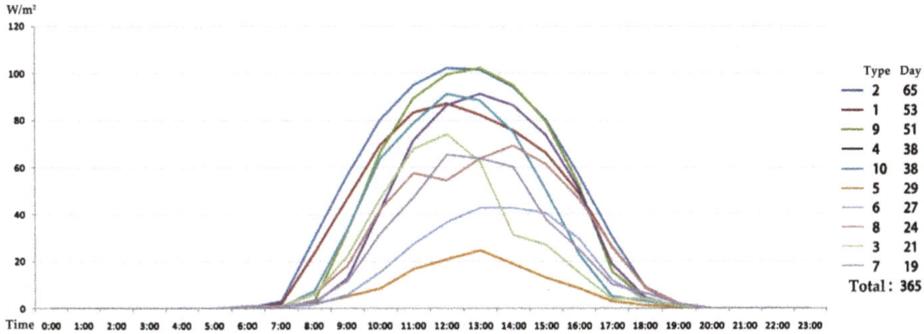

Figure 10. Classification results of annual hourly photovoltaic output changes per unit area of horizontal plane (clustering center). Source: Drawn by the author.

Categories 5, 6, 8, 3, and 7 are heavily affected by weather factors such as overcast rain, and their hourly photovoltaic power generation decreases in different periods. On the other hand, categories 2, 1, 9, 4, and 10 represent typical hourly photovoltaic power generation conditions on sunny days, with power generation being less affected by the weather. The average value of PV power output per unit area at each time point of these five types of curves is used as the standard hourly PV power output per unit area per day curve developed in this paper to reflect the local situation throughout the year (refer to Figure 11 for the results). Then, based on the installed capacity per unit area and the available PV area of the roof, facade, and ground parking area of each block, the total installed PV capacity, the total hourly PV power generation, and the hourly PV power generation density of each block can be calculated, as shown in Equations (20)–(22).

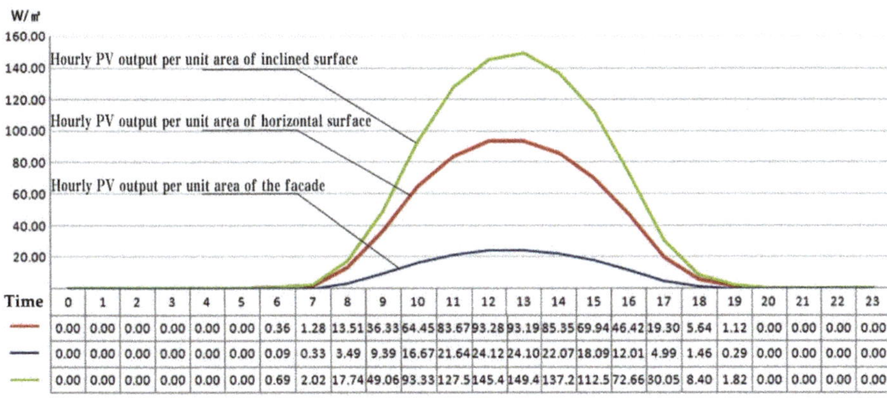

Time	0	1	2	3	4	5	6	7	8	9	10	11	12	13	14	15	16	17	18	19	20	21	22	23
Hourly PV output per unit area of horizontal surface (W/㎡)	0.00	0.00	0.00	0.00	0.00	0.00	0.36	1.28	13.51	36.33	64.45	83.67	93.28	93.19	85.35	69.94	46.42	19.30	5.64	1.12	0.00	0.00	0.00	0.00
Hourly PV output per unit area of the facade (W/㎡)	0.00	0.00	0.00	0.00	0.00	0.00	0.09	0.33	3.49	9.39	16.67	21.64	24.12	24.10	22.07	18.09	12.01	4.99	1.46	0.29	0.00	0.00	0.00	0.00
Hourly PV output per unit area of inclined surface (W/㎡)	0.00	0.00	0.00	0.00	0.00	0.00	0.69	2.02	17.74	49.06	93.33	127.5	145.4	149.4	137.2	112.5	72.66	30.05	8.40	1.82	0.00	0.00	0.00	0.00

Figure 11. Hourly output power curve of standard daily PV per unit area used in this paper. Source: Drawn by the author.

The assessment of total installed PV capacity of each block is performed as follows:

$$C = Sr \cdot Cph + Sf \cdot Cpv + St \cdot Cph \tag{20}$$

The assessment of hourly photovoltaic power generation in various blocks is performed as follows:

$$PV_i = \frac{Sr \cdot Oph_i + Sf \cdot Opv_i + St \cdot Oph_i}{1000} \quad (21)$$

The assessment of hourly photovoltaic power generation density in various blocks is performed as follows:

$$PVd_i = \frac{1000 \cdot PV_i}{A} \quad (22)$$

C (W) is the total installed photovoltaic capacity of a certain block. Cph (W/m^2) is the estimated installed capacity per unit area of the horizontal plane of the proposed area. Cpv (W/m^2) is the estimated installed capacity per unit area of the proposed area façade.

Sr (m^2) is the available rooftop photovoltaic area. Sf (m^2) is the photovoltaic available area for a block building facade (south). St (m^2) is the photovoltaic available area of a block of ground parking area. PV_i (kWh) is the photovoltaic power generation of a certain area at the ith moment. Oph_i (W/m^2) is the photovoltaic output power per unit area of the horizontal plane at the ith time. Opv_i (W/m^2) is the photovoltaic output power per unit area of the facade at the ith time. PVd_i (Wh/m^2) is the photovoltaic power generation density at the ith moment.

3.2. *Hourly Electricity Load Calculation*

3.2.1. Load Density Index Method

In this paper, the power load forecasting method recommended in the Urban Power Planning Code [39], known as the load density index method (also referred to as the power per unit area method), is utilized to depict the electricity consumption of different buildings in the city. According to a survey conducted in cities of various sizes, focusing on urban planning and development, population planning, and the growth of residents' income levels, this method utilizes the general peak load power per unit area (W/m^2 or kW/km^2), known as the load density index [40], to assess the urban load level. This provides a reference for the power supply capacity configuration in this area. At the same time, since this method directly reflects the power load situation by the product of the building area or regional occupation area and load density index (Equation (23)), it is easier to effectively combine with the GIS geographic information system in terms of operation. This allows for a quick realization of the assessment of power load for various urban buildings.

In order to conduct hourly load assessments of different buildings in the city, this paper introduces the concept of hourly load rate (Lr_i) and the daily hourly load rate curve, drawing on the idea of power grid load rate. This approach aims to illustrate the variations in hourly load magnitude across different buildings. Finally, this paper provides a brief evaluation of the hourly power load of blocks in Hohhot City using two indices: load density and hourly load rate, as shown in Equation (24).

$$P_C = Ld \cdot A_{built} = Ld \cdot FAR \cdot A \quad (23)$$

$$Load_i = Lr_i \cdot Ld \cdot A_{built} \quad (24)$$

P_C (kW) is the peak load power (active power) of each building. Ld (W/m^2) is the load density of each building. A_{built} (m^2) is the building area. FAR is the floor area ratio of each block. A (m^2) is the land area of a certain place. $Load_i$ (kWh) is the power load at the ith moment of each block. Lr_i is the load rate at the ith moment of each block.

3.2.2. Load Density Index

The load density index is typically utilized to predict the power supply capacity configuration of different blocks, and its value also indicates the peak load of a block. In this paper, the proposed values of the load density index for various land uses are shown in Table 4.

Table 4. Partial land load density indicators selected in this paper [41–43].

Nature of the Land	Coding	Average Load Density <Ld> (W/m^2)	Simultaneous Factor (Kt)	Demand Factor (Kd)	Proposed Load Density Ld (W/m^2)
Residential land	R	55.00	0.95	0.43	22.09
Public administration and public service	A	74.44	0.98	0.75	54.44
Retail commercial land	B11	96.00	0.98	0.85	79.56
Wholesale market land	B12	62.50	0.98	0.85	51.80
Land for catering industry	B13	100.00	0.98	0.85	82.88
Land for the hotel	B14	70.00	0.98	0.85	58.01
Commercial office land	B2	81.00	0.98	0.85	67.13
Category I industry	M1	53.71	0.97	0.35	18.24
Category II industry	M2	59.43	0.97	0.38	21.62
Category III Industry	M3	74.00	0.97	0.43	30.51
Logistics and storage land	W	24.60	0.95	0.38	8.76
Municipal utility facilities	U	42.20	0.95	0.38	15.03
Green land and square land	G	2.00	0.95	0.75	1.43

Data Source: collated and calculated by the author according to relevant literature.

3.2.3. Daily Hourly Load Rate Curve

Analyzing the electricity load curves of various sections provided by Hohhot National Grid, we utilized SPSS software to classify the daily hourly load rate curves of 290 land samples based on the land use characteristics. Subsequently, we analyzed the clustering results to extract the most representative daily load rate curves of different land use types. These representative curves were then considered the typical daily load rate curves proposed in this paper (Figure 12).

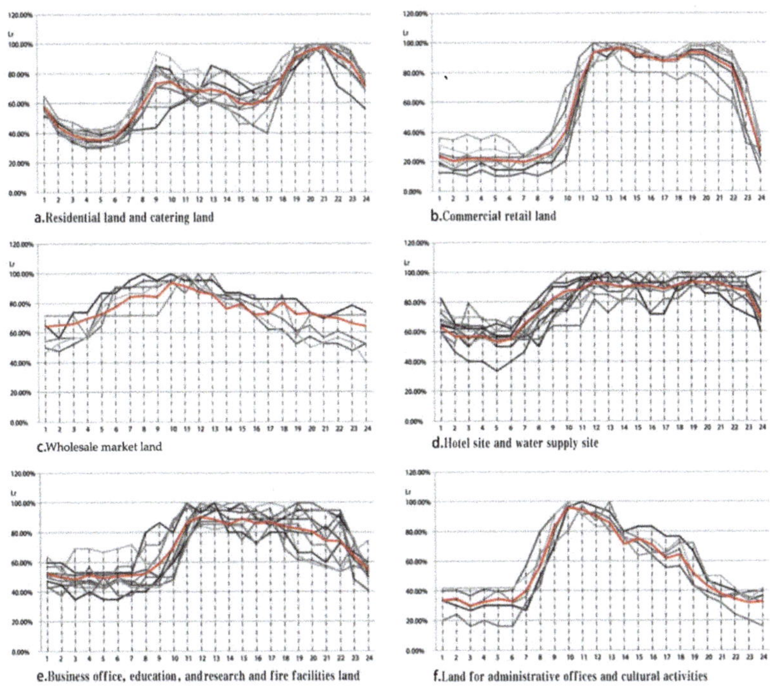

Figure 12. *Cont.*

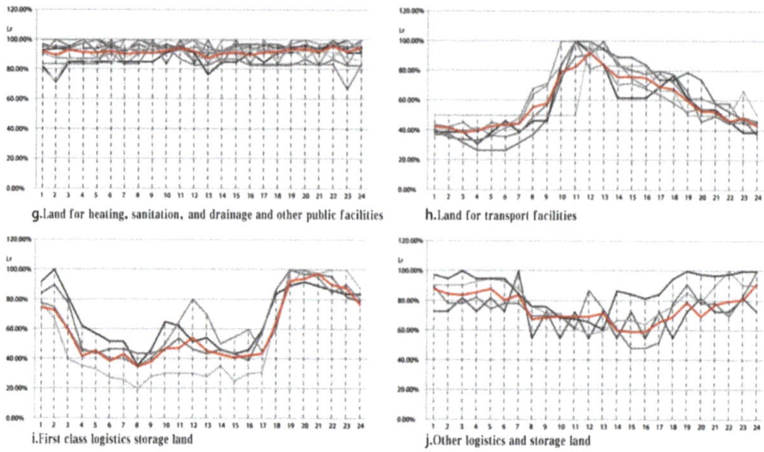

Figure 12. Daily load rate curve of each civil building block proposed in this paper (red): (**a**) Residential land and catering land; (**b**) Commercial retail land; (**c**) Wholesale market land; (**d**) Hotel site and water supply site; (**e**) Business office, education, and research and fire facilities land; (**f**) Land for administrative offices and cultural activities; (**g**) Land for heating, sanitation, and drainage and other public facilities; (**h**) Land for transport facilities; (**i**) First class logistics storage land; (**j**) Other logistics and storage land. Data source: Drawn by the author according to the results of SPSS cluster analysis.

3.2.4. Hourly Load Assessment for Mixed-Function Plots

In addition to the properties of all kinds of single-function land use, there are also various mixed-use plots in the city. These plots do not have a dominant single building function; instead, they showcase a variety of building functions coexisting. The mixed-function plot mainly combines horizontal and comprehensive functions. On a smaller scale (single building), it is primarily a combination of vertical and temporal dimensions. On a block scale, all four hybrids can be achieved.

Therefore, this paper selected 12 mixed commercial and other functional plots in Hohhot City and analyzed the proportion of the floor area of their respective commercial parts (Figure 13). The results show that most businesses account for between 20% and 40%. In order to facilitate the assessment of potential, this paper adopts the proportion of 30% commercial buildings as the standardized metric for all mixed-use plots. This means that the ratio of commercial to other building functional areas is 3:7. The peak load density and hourly daily load rate of all mixed-use plots are then calculated based on this standard. The peak load density values are B11/R: 39.67 W/m^2, B11/B14: 61.47 W/m^2, B11/B2: 65.72 W/m^2.

It is important to note that this article has established an ideal standard curve for photovoltaic power generation, depicting the hourly output per unit area. The results are derived from a year-long simulation of photovoltaic hourly output over 365 days. The analysis includes the calculation of the similarity to the average output power over 245 days, specifically focusing on the period from 6 to 7 p.m. Therefore, in the three indicators of daytime photovoltaic replacement rate (ESRd), photovoltaic utilization rate (PUR), and standard deviation of hourly replacement rate (Stdev), this paper uses a 13-hour period as the standard for calculation. The hourly PV power generation of each block is calculated based on the assessment results of the available PV area of each block and the proposed hourly PV power output curve per unit area of the standard daily area (Figure 12). The hourly power load of each block is determined by the proposed curve of hourly load rate for each functional building and the proposed index of load density for each functional building (Table 4), which is calculated using Equation (24). Then, according to Equations (1)–(3), the three indicators of ESRd, PUR, and Stdev of each block can be calculated, as shown in Figures 14–16.

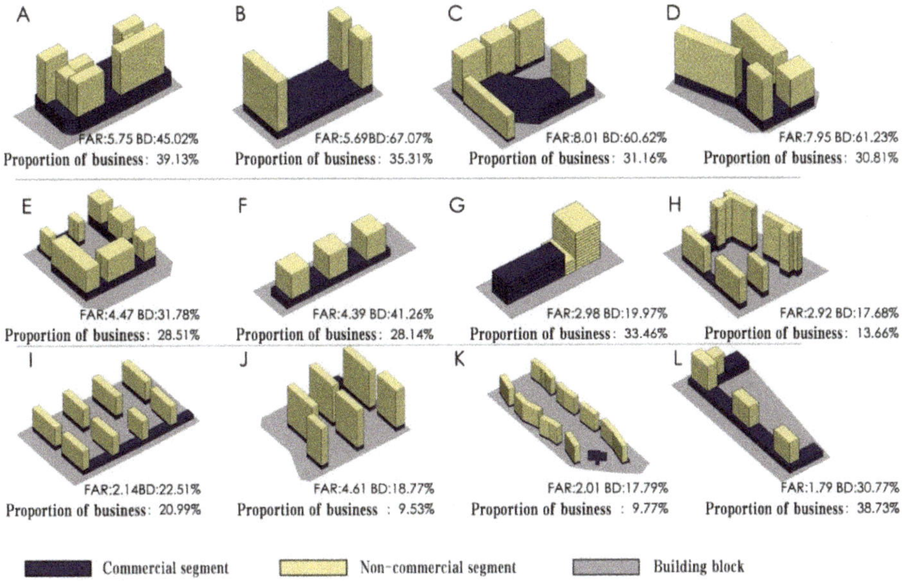

Figure 13. (**A–L**) Commercial area proportion of 12 commercial and other functional mixed plots selected in this paper: Source: Drawn by the author.

Figure 14. Daytime photovoltaic replacement rate (ESRd) of blocks in Hohhot.

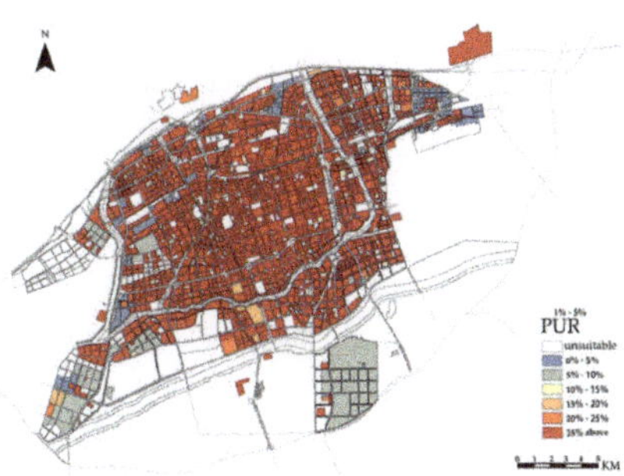

Figure 15. Photovoltaic utilization rate of various blocks in Hohhot (PUR).

Figure 16. Standard deviation of inter-day displacement rate (Stdev) of blocks in Hohhot.

From the perspective of PV installation potential, the total area of suitable land for PV installation in all types of construction land in Hohhot is 1898.18 million square meters, accounting for 57.6% of the total construction land area. The total available rooftop PV area is 604,447,647 square meters, with an installed capacity of 95,507,028 kW. The total available PV area on the south elevation is 249,144,454 square meters, with an installed capacity of 132,046 kW. The total available PV area in the ground parking area is 49,499,580 square meters, with a total installed PV capacity of 78,203,034 kW. The total installed PV capacity in the city is about 10,464 MW. According to the typical daily hourly PV power generation, the theoretical average daily power generation can reach 44,015,000 kWh when tilt installation is adopted, accounting for approximately 73.7% of the daily electricity consumption of the entire society.

In terms of total photovoltaic power generation, Hohhot city presents optimistic development prospects due to its excellent solar radiation resources. In this paper, the average daytime PV replacement rate for all types of industrial sites is 82.9%, based on the daytime PV replacement rate (ESRd). This reflects the application of PV to replace traditional energy sources during off-peak production periods in all types of plants, even when there is excess power output. After considering the hourly electricity usage per unit of land for PV generation, the actual daily shipments in the city amount to 31,281 million kWh, representing approximately 52.4% of the total electricity consumption (citywide PV replacement rate). In other words, around 28.9% of the PV generation is theoretically wasted due to mis-match with the electricity load, which is equivalent to an indirect loss of 12,734 million kWh of electricity output per day.

The stability of hourly photovoltaic power supply is determined by the standard deviation of the hourly photovoltaic replacement rate (Stdev). Spatial quantification of photovoltaic energy using urban data is becoming an important element of future smart cities and sustainable development. On the whole, the average Stdev of the city is 23.8%, which indicates relative stability in terms of the overall variability within the city. The average Stdev level of all types of industrial land is the highest at 38.9%, indicating significant instability. Residential and office space occupancy rates are relatively stable at 21.8% and 15.9%, respectively. The most stable type of land is commercial land, with an average Stdev of 8.8%. The main reason is that the land has a high load density, and the impact of photovoltaic power is not significant. Even after the implementation of photovoltaic systems, urban distribution networks still need to maintain a high level of capacity allocation and power supply to ensure reliable electricity distribution.

3.3. Cluster Analysis of Power Characteristic Laws of Different Plots

In order to summarize the evolving patterns of daily power characteristics following the implementation of PV power generation at various sites, this study conducted a cluster analysis of 5023 plots in the city using three indicators: daytime PV replacement rate, PV utilization rate, and PV utilization stability (Table 5). Additionally, a regression analysis was carried out to examine the relationship between each indicator and the floor area ratio (FAR) (Figure 17).

Table 5. Cluster analysis of daily power characteristics after PV installation.

Category	Center of Clustering			Major Building Features and Their Corresponding Floor Area Ratio Ranges
	ESRd/%	**PUR/%**	**Stdev/%**	
1	0.00	0.00	0.00	Cultural relics, monuments, religious facilities, special sites, various types of plazas, green spaces, water bodies, and other areas where photovoltaic installation is not feasible
2	88.52	1.20	24.79	Various types of social car parks
3	79.67	5.12	37.01	Primary and secondary schools (0.8–1.19), logistics warehouses (0.8)
4	79.25	11.43	40.27	Executive offices (0.8–0.99), primary and secondary schools (1.2–2.2), various industrial buildings (0.8), integrated traffic hubs (0.86–1.49)
5	52.32	26.58	40.55	Executive offices (1.11–2.13), educational research (0.84–2.1), medical and health (0.81–1.47), residences (0.85–1.78), mixed residential/commercial (0.82–1.04), integrated traffic hubs (1.71–2.58)
6	29.31	28.66	25.70	Executive offices (2.19–3.21), educational research (2.12–3.82), medical and health (2.22–3.57), commerce (0.96–1.31), mixed commercial/offices (0.97–1.29), business offices (0.83–1.18), residences (1.82–2.94), mixed residential/commercial (1.44–1.84)
7	19.18	28.25	15.67	Executive offices (3.53–4.33), educational research (3.8–4.15), medical and health (3.65–4.38), commerce (1.42–1.76), mixed commercial/offices (1.34–2.32), business offices (1.2–2.1), residences (2.95–4.39), mixed residential/commercial (1.89–3.4)
8	10.73	28.80	7.59	Executive offices (5.92–5.98), educational research (4.24–4.38), commerce (1.93–3.8), mixed commercial/hotels (2.53–5.74), mixed commercial/offices (2.34–6.0), business offices (2.19–5.64), residences (5.74–5.94), mixed residential/commercial (3.32–5.81)

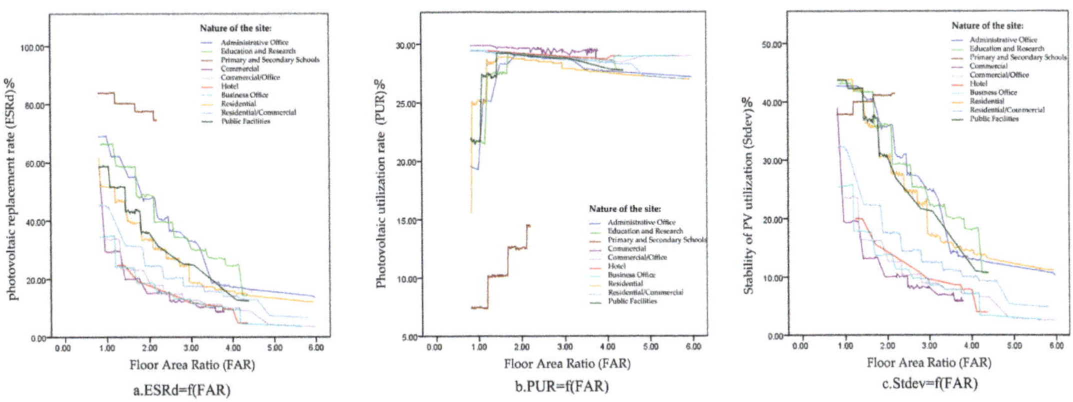

Figure 17. Regression analysis of ESRd = f(FAR), PUR = f(FAR), Stdev = f(FAR).

After considering the time sequence of PV power consumption, all urban land plots can be categorized into three main types: PV unsuitable land plots (category 1), PV excess power land plots (categories 2 to 4), and general PV application land plots (categories 5 to 8). Among them, the areas with excess PV power mainly include industrial areas, warehouses, car parks, primary and secondary schools, and other building plots with insufficient power consumption. The replacement rate of PV systems is excessively high, and the surplus power cannot be consumed during the peak hours of PV power generation. As a result, the PV utilization rate is very low, and the replacement rate fluctuates significantly throughout the day (categories 3 and 4). The general PV application areas mainly include commercial, office, residential, and other civil buildings. Based on the results of the evaluation and analysis (Table 5, Figure 17), it is evident that the variation in electricity load demand due to differences in plot ratio will significantly influence the electricity consumption patterns following the installation of photovoltaic systems in each plot. Therefore, it is of great significance to enhance the overall PV potential of the city by effectively controlling the planning layout and plot ratio of each site.

As far as the stage of PV development is concerned, traditional fossil energy generation, mainly thermal power, will continue to dominate as a reliable energy source for an extended period. However, the current thermal turbines are generally characterized by slow start-up and shutdown, high minimum technical output requirements, lack of flexibility in system regulation, and rigidity in power scheduling. These limitations may pose challenges in meeting the future demands of renewable energy development. Considering its development goals, this paper takes the 40% daytime PV replacement rate as the best criterion to measure the proportion of PV power consumption in each urban area. During the daytime when PV energy is available, 40% of the load demand is supplied by PV in combination with a small amount of energy storage to ensure voltage stability. The remaining load demand is met by 40% of the minimum thermal power output and 20% from other renewable energy sources. As can be seen in Table 4, the daytime PV replacement rate in the areas of plot 5 and plot 6 has a better match with this criterion, with ESRd values of 52.32% and 29.31%, respectively. Meanwhile, it can be seen from Figure 17 that under different plot ratios, there is a theoretical maximum peak value of the PV utilization rate for each plot. However, due to the relatively low plot ratio of category 5, its economic return from PV systems is poor. At the same time, the stability of the power supply also reflects a significant disadvantage, with a Stdev value as high as 40.55%. This high variability is relatively unfavorable for regulating the thermal system, given the substantial demand of the grid connection. After a comprehensive comparison, the building plot ratio interval for each type of building will serve as the standard for the sixth type of land, which will be more favorable to the overall PV potential of Hohhot. The corresponding values of the functional plot ratios for each type of building in this category will become the primary focus for adjusting urban planning in the future. It can be further seen that various types of small high-rise (average number of floors 6–9) building forms in the city may be better suited to meet the needs of urban PV utilization and will likely become the primary spatial layout for future urban development. These building forms include residential buildings, administrative office buildings, and more.

In addition, as shown in Table 5, the PV replacement rate is high for all types of industrial land (Class 3 and 4) in the city, indicating a clear surplus of electricity output. The paper examines the impact of the demand coefficient, which indicates that not all power equipment is activated simultaneously. The power load is only represented by daily office, lighting, and other non-production loads. In fact, various types of industrial buildings require significant power loads during peak production hours. Typically, these power needs are met by specific power plants through dedicated lines to ensure a consistent power supply. However, photovoltaic power faces challenges in meeting these demands due to its inherent instability and intermittency characteristics. Hence, industrial plants situated far from other urban functions and centrally located in isolated patches may experience minimal actual benefits from PV applications.

In addition, due to the higher load density in commercial, hotel, business office, and other land areas in the city (categories 7, 8), the overall rate of photovoltaic replacement is low. According to this paper, various types of building functions need to adjust the volume ratio interval. The average volume ratio needs to be adjusted downward to 0.96–1.31 (for instance, in commercial buildings, the average number of floors is typically 2–3). Otherwise, the shape of the building will exceed the reasonable design range and be challenging to

realize. Therefore, it is difficult to realize the future urban development of the above area. Therefore, for future adjustments to urban planning in the aforementioned areas, more consideration should be given to facilitating the layout of neighborhoods or even the organic integration of these areas while ensuring that the normal operation of the city is not disrupted.

4. Conclusions

The strategic deployment of installed PV systems within built environments is paramount for driving future urban renewable energy structural transformations. Simply focusing on guiding PV planning and construction based on total power generation at a singular level cannot fully exploit the potential benefits of this approach. The urban distributed photovoltaic energy evaluation model introduced in this paper offers a more comprehensive integration of future energy needs with sustainable urban development. It lays the groundwork for incorporating Internet of Things and big data technologies, which are crucial for maximizing future urban energy potential. However, as this paper solely addresses the optimization of PV potential in Hohhot city concerning a single spatial layout issue of volume ratio adjustment, several limitations remain regarding parameter selection, data accuracy, and influencing factors. Future research should encompass a broader spectrum of practical social development issues, such as economic growth, urban planning, and industrial adaptations. This holistic approach will facilitate strategic planning of urban energy spatial layouts, fostering the transformation of photovoltaic power energy structures and advancing sustainable urbanization.

Author Contributions: Conceptualization, Z.Z.; methodology, S.C. and C.W.; software, S.C.; validation, L.T., H.L. and R.H.; date analysis, S.C. and Z.Z.; investigation, S.C. and C.W.; resources, S.C. and R.Z.; data curation, H.L. and H.Y.; writing-original draft preparation, S.C.; writing-review and editing, H.L., H.Y. and R.H.; visualization, Z.Z. and R.H.; supervision, C.W., H.Y. and R.Z.; project administration, L.T.; funding acquisition, L.T. and R.Z. All authors have read and agreed to the published version of the manuscript.

Funding: This work was supported by the National Natural Science Foundation of China (52078324); the National Natural Science Foundation of China (52078322); Tianjin University Graduate Students' Award Program for Outstanding Innovation in Arts and Sciences (A1-2022-005) and Tianjin Graduate Student Research Innovation Program (2022BKY093).

Data Availability Statement: The raw data supporting the conclusions of this article will be made available by the authors on request.

Conflicts of Interest: Author Siyuan Chen was employed by the company China Northeast Architectural Design & Research Institute Co., Ltd. The remaining authors declare that the research was conducted in the absence of any commercial or financial relationships that could be construed as a potential conflict of interest.

Nomenclature

Ar (m^2)	Roof area/floor area
Ap (m^2)	Area of selected PV module
A_{built} (m^2)	Building area
β	Best angle (37.78° in Hohhot)
BD′	Building density
BD	Proposed building density
Cph (W/m^2)	Installed capacity per unit area of the horizontal plane
Cpv (W/m^2)	Installed capacity per unit area of the facade
C (W)	Installed photovoltaic capacity
ESR_i	Hourly Energy Substitution Rate
ESR	Overall Energy Substitution Rate
FAR	Floor area ratio

F′	Average number of building floors
F/f	Proposed number of building floors
G	Green land rate
H (m)	Building height
Kf	Window reduction coefficient
Ks	Facade and shadow reduction coefficient
Kr	Rooftop photovoltaic utilization coefficient
Ld (W/m^2)	Load density of each building
$Load_i$ (kWh)	Power load at the ith moment
Lr_i	Load rate at the ith moment
L0	Reference sunshine spacing coefficient
N	Number of power generation hours
Oph_i (W/m^2)	Photovoltaic output power per unit area of the horizontal plane at the ith time
Opv_i (W/m^2)	Photovoltaic output power per unit area of the facade at the ith time
PUR	Photovoltaic utilization rate
Pw (Wp)	Peak power (rated power)
Pr	PV coverage rate
PV_i (kWh)	Photovoltaic power generation at the ith moment
PVd_i (Wh/m^2)	Photovoltaic power generation density at the ith moment
Pc (kW)	Peak load power (active power)
P	Percentage of ground parking land
Pe_i (kWh)	Consumption of photovoltaic power
PV_i (kWh)	Photovoltaic power generation
R	Land use rate
R_i	Arithmetic average of daily hourly PV replacement rate
Stdev	Standard deviation of hourly PV replacement rate
Sr (m^2)	Available rooftop photovoltaic area
Sf (m^2)	Available area for photovoltaic of a block building facade (south)
St (m^2)	Photovoltaic available area of a block of ground parking area
Sn (m^2)	Statistical value of the south elevation area
Sf	Photovoltaic available area
S	Rate of supporting public construction land
T (m)	Building depth
δ	Sun azimuth

References

1. UN-Habitat. *World Cities Report 2020: The Value of Sustainable Urbanization*; United Nations Human Settlements Programme: Nairobi, Kenya, 2020.
2. Kartal, G. The Effects of Positive and Negative Shocks in Energy Security on Economic Growth: Evidence from Asymmetric Causality Analysis for Turkey. *Econ. Comput. Econ. Cybern. Stud. Res.* **2022**, *56*, 223–240.
3. Yoon, J.; Joung, S. Environmental self-identity and purchasing ecofriendly products. *J. Logist. Inform. Serv. Sci.* **2021**, *8*, 82–99.
4. IRENA. *Future of Solar Photovoltaic: Deployment, Investment, Technology, Grid Integration and Socio-Economic Aspects (A Global Energy Transformation: Paper)*; International Renewable Energy Agency: Abu Dhabi, United Arab Emirates, 2019.
5. Root, L.; Perez, R. *Photovoltaic Covered Parking Lots: A Survey of Deployable Space in the Hudson River Valley, New York City, and Long Island, New York*; Atmospheric Sciences Research Center: Albany, NY, USA, 2006.
6. Xiong, Y.; Liu, X.; Ma, S. Analysis on application conditions of urban scale photovoltaic power generation in China. *Renew. Energy Resour.* **2012**, *30*, 123–126.
7. Manni, M.; Formolli, M.; Boccalatte, A.; Croce, S.; Desthieux, G.; Hachem-Vermette, C.; Kanters, J.; Ménézo, C.; Snow, M.; Thebault, M.; et al. Ten questions concerning planning and design strategies for solar neighborhoods. *Build. Environ.* **2023**, *246*, 110946. [CrossRef]
8. IEA. *PVPS Task VII, Potential for Building Integrated Photovoltaics*; International Energy Agency: Paris, France, 2002.
9. Izquierdo, S.; Rodrigues, M.; Fueyo, N. A Method for Estimating the Geographical Distribution of the Available Roof Surface Area for Large-Scale Photovoltaic Energy-Potential Evaluations. *Sol. Energy* **2008**, *82*, 929–939. [CrossRef]
10. Jouri, K. *Planning for Solar Buildings in Urban Environments an Analysis of the Design Process, Methods and Tools*; Department of Architecture and Built Environment, Faculty of Engineering LTH, Lund University: Lund, Sweden, 2015.
11. Zhao, B.; Xiao, C.; Xu, C. Analysis of absorptive capacity of distributed photovoltaic grid connection in regional distribution network based on permeability. *Electr. Power Syst.* **2017**, *21*, 111–117.

12. Cheng, M.; Zhao, S.; Wu, L. Comprehensive evaluation of urban photovoltaic power generation projects based on AHP-TOPSIS. *Electr. Power* **2016**, *49*, 171–175.
13. Sigrid Lindner of Ecofys. Solar Urban Planning Berlin [EB/OL]. Available online: http://www.pvupscale.org/IMG/pdf/Berlin.pdf (accessed on 10 May 2020).
14. Zhang, H. *Study on the Evaluation of Photovoltaic Utilization Potential of Urban Building Roofs*; Tianjin University: Tianjin, China, 2017.
15. Enongene, K.E.; Abanda, E.H.; Otene, I.J.J.; Obi, S.I.; Okafor, C. The potential of solar photovoltaic systems for residential homes in Lagos city of Nigeria. *J. Environ. Manag.* **2019**, *244*, 247–256. [CrossRef] [PubMed]
16. Rosas-Flores, J.A.; Zenon-Olvera, E.; Galvez, D.M. Potential energy saving in urban and rural households of Mexico with solar photovoltaic system using geographical information system. *Renew. Sustain. Energy Rev.* **2019**, *116*, 109412. [CrossRef]
17. Bodis, K.; Kougias, I.; Jager-Waldau, A.; Taylor, N.; Szabo, S. A high-resolution geospatial assessment of the rooftop solar photovoltaic potential in the European Union. *Renew. Sustain. Energy Rev.* **2019**, *114*, 109309. [CrossRef]
18. Verso, A.; Martin, A.; Amador, J.; Dominguez, J. GIS-based method to evaluate the photovoltaic potential in the urban environments: The particular case of Miraflores de la Sierra. *Sol. Energy* **2015**, *117*, 236–245. [CrossRef]
19. Saadaoui, H.; Ghennioui, A.; Ikken, B.; Rhinane, H.; Maanan, M. Using GIS and photogrammetry for assessing solar photovoltaic potential on Flat Roofs in urban area case of the city of Ben Guerir/Morocco. In Proceedings of the 5th International Conference on Geoinformation Science—GeoAdvances 2018, Casablanca, Morocco, 10–11 October 2018; Copernicus Publications: Göttingen, Germany, 2019; Volume XLII-4/W12, pp. 155–166.
20. Wang, L.Z.; Tan, H.W.; Ji, L.; Wang, L. A method for evaluating photovoltaic potential in China based on GIS platform[C]//IOP Conference Series: Earth and Environmental Science. *IOP Publ.* **2017**, *93*, 012056.
21. Kozlovas, P.; Gudzius, S.; Ciurlionis, J.; Jonaitis, A.; Konstantinaviciute, I.; Bobinaite, V. Assessment of technical and economic potential of urban rooftop solar photovoltaic systems in Lithuania. *Energies* **2023**, *16*, 5410. [CrossRef]
22. Alhamwi, A.; Medjroubi, W.; Vogt, T.; Agert, C. Development of a GIS-based platform for the allocation and optimisation of distributed storage in urban energy systems. *Appl. Energy* **2019**, *251*, 113360. [CrossRef]
23. Tapia, M.; Ramos, L.; Heinemann, D.; Zondervan, E. Power to the city: Assessing the rooftop solar photovoltaic potential in multiple cities of Ecuador. *Phys. Sci. Rev.* **2023**, *8*, 2285–2319. [CrossRef]
24. Liu, J.; Wu, Q.; Lin, Z.; Shi, H.; Wen, S.; Wu, Q.; Zhang, J.; Peng, C. A novel approach for assessing rooftop-and-facade solar photovoltaic potential in rural areas using three-dimensional (3D) building models constructed with GIS. *Energy* **2023**, *282*, 128920. [CrossRef]
25. Bernasconi, D.; Guariso, G. Rooftop PV: Potential and Impacts in a Complex Territory. *Energies* **2021**, *14*, 3687. [CrossRef]
26. Fu, R.; Timothy, R.; Robert, M. *2018 U.S. Utility—Scale Photovoltaics-Plus-Energy Storage System Costs Benchmark*; National Renewable Energy Laboratory: Oak Ridge, TN, USA, 2018.
27. Inner Mongolia Hohhot Bureau of Statistics. *2023 Hohhot City Statistical Yearbook [EB]*; Hohhot Bureau of Statistics: Huhehaote, China, 2023.
28. Andson, K.; Coddington, M.; Burman, K. *Interconnecting PV on New York City's Secondary Network Distribution System*; National Renwable Energy Laboratory: Golden, MO, USA, 2009.
29. Hua, Z. *Research on the Assessment of the Potential of Urban Building Rooftop Photovoltaic Utilization*; Tianjin University: Tianjin, China, 2017.
30. Ren, B. *Research on Low Energy Consumption Design Prototype of Multi-Storey Office Buildings in Cold Regions*; Tianjin University: Tianjin, China, 2014.
31. Hang, Z. *Zhejiang Province of Planning Bureau, Rules for Building Storey Height Control and Floor Area Ratio Index Calculation*; Hangzhou Municipal Planning Bureau: Hangzhou, China, 2010.
32. *GB50099-2011*; Ministry of Housing and Urban-Rural Construction of the People's Republic of China. Code for Design of Primary and Secondary Schools. Guangming Daily Press: Beijing, China, 2011.
33. *GB 51039-2014*; National Health and Family Planning Commission of the People's Republic of China. Code for Design of Polyclinic Buildings. China Planning Press: Beijing, China, 2014.
34. *GB50189-2015*; China Academy of Building Sciences. Design Standard for Energy Efficiency of Public Buildings. China Building Industry Press: Beijing, China, 2015.
35. *GB50797-2012*; China Electricity Council. Code for Design of Photovoltaic Power Station. China Planning Press: Beijing, China, 2012.
36. *GB50180-93*; Ministry of Housing and Urban Rural Development of the People's Republic of China. Code for Planning and Design of Urban Residential Areas. China Building Industry Press: Beijing, China, 2002.
37. Brito, M.; Freitas, S.; Guimar, E. The importance of facades for the solar PV potential of a Mediterranean city using LiDAR data. *Renew. Energy* **2017**, *111*, 85–94. [CrossRef]
38. *JGJ203-2010*; Ministry of Housing and Urban Rural Development of the People's Republic of China. Technical Code for Application of Solar PV System in Civil Buildings. China Construction Industry Press: Beijing, China, 2010.
39. *GB/T50293-2014*; Ministry of Housing and Urban Rural Development of the People's Republic of China. Code for Urban Power Planning. China Construction Industry Press: Beijing, China, 2014.
40. The General Institute of Electric Power Planning and Design. *Ministry of Electric Power Industry, Power System Design Manual*; China Electric Power Press: Beijing, China, 1998.

41. China Institute of Building Standards Design. *Technical Measures for National Civil Building Engineering Design: 2009 Edition, Electrical*; China Planning Press: Beijing, China, 2009.
42. Ye, X. *Research on Power Load Index of Ningbo Unit Construction Land, Urban Times, Collaborative Planning—2013 China Urban Planning Annual Conference*; China Urban Planning Society: Beijing, China, 2013.
43. Liu, D.; Zhu, Y.; Zhang, Q. Discussion on the Value of Load Index, Demand Coefficient and Synchronization Coefficient in Urban Power Planning. *Archit. Eng. Technol. Des.* **2014**, *10*, 16.

buildings

MDPI

Article

Revitalizing Rural Landscapes: Applying Cultural Landscape Gene Theory for Sustainable Spatial Planning in Linpu Village

Shuhu Liu [1], Lin Wu [1], Chenlu Xiang [1] and Wenyuan Dai [2,*]

[1] Institute of Urban and Rural Planning, School of Architecture and Town and Country Planning, Fuzhou University, Fuzhou 350108, China; liushuhu@fzu.edu.cn (S.L.); 231520024@fzu.edu.cn (L.W.); 231520027@fzu.edu.cn (C.X.)

[2] Institute of Geography, School of Geographical Sciences, Fujian Normal University, Fuzhou 350117, China

* Correspondence: dwygeo@fjnu.edu.cn

Abstract: In the context of rapid urbanization, it is imperative to address the preservation and sustainable development of rural landscapes. This study introduces a novel framework using cultural landscape gene theory to enhance the landscape spatial pattern of Linpu Village, China. Employing a multi-method approach, including data collection and historical data analysis, GIS analysis, and field surveys, we analyze the village's landscape features and identify a well-preserved "one core, two wings" spatial configuration with diminishing accessibility from the center to the periphery. The study uncovers significant challenges to landscape coherence and legibility, including landscape element discontinuity, inadequate maintenance, and fragmented information. To address these issues, we propose optimization strategies aimed at revitalizing weakened cultural landscape genes, repairing disrupted elements, ensuring corridor integrity, and improving spatial organization. The extraction–construction–analysis–preservation model established in this research not only delivers a comprehensive analysis of Linpu's cultural landscape but also offers a scalable framework for preserving the ecological and cultural integrity of traditional villages. This study contributes to the practical implementation of sustainable landscape strategies, enhancing the ecological adaptability and resilience of rural environments in the face of urban encroachment.

Keywords: cultural landscape; sustainable development; human environment; livelihoods

1. Introduction

Traditional villages are vital custodians of cultural heritage, encapsulating the historical and architectural progression of societies worldwide [1]. These villages—from the ancient towns of Europe to the heritage settlements in Africa and the rural landscapes in Asia—are under increasing threat from rapid urbanization [2–4]. This phenomenon leads to the degradation of cultural landscapes and disrupts historical continuity, posing significant challenges to the sustainable development of traditional villages globally [5].

Historically, global preservation efforts have predominantly concentrated on tangible elements, such as architectural styles and the structural conservation of buildings [6]. However, these efforts often neglect the broader landscape structures and cultural contexts integral to the villages' identities [7]. This oversight can result in fragmented and sometimes ineffective conservation practices. Therefore, it is crucial to adopt a holistic approach that encompasses both the physical and cultural dimensions of these landscapes.

Cultural Landscape Genes of Traditional Settlements (CLGTS) provides a robust framework for analyzing and preserving the essence of traditional village landscapes [8]. This theory, which conceptualizes 'Cultural Landscape Genes' as the cultural, physical, and historical narratives embedded within traditional settlements, offers a way to maintain the identity and continuity of these landscapes [9]. By treating these 'genes'—comprising physical features like buildings and roads, cultural practices, and historical narratives—as components that define a landscape's character and resilience, landscape gene theory enables a nuanced conservation strategy that preserves regional cultural attributes in an integrated and sustainable manner.

Despite advancements, cultural landscape gene theory faces critical research gaps. Firstly, its application in rural planning remains limited, with a significant need to integrate this theoretical framework into practical, scalable conservation strategies [10]. Secondly, the predominant qualitative nature of current analysis methods highlights a pressing need for systematic quantitative approaches [11]. Thirdly, the application of this research has predominantly been confined within China, specifically to traditional villages in provinces such as Hunan and Shaanxi [12,13], resulting in geographical concentration and limiting broader applicability.

This study addresses these gaps by employing Linpu Village in Fuzhou, in southeastern China, as a case study to develop a sustainable model for traditional village landscapes using landscape gene theory. This research has three objectives. The first is to establish an "extraction-construction-analysis-preservation" model to enhance the practical application of the theory in rural planning. The second is to enhance analytical methods using Geographic Information Systems (GIS) to analyze key spatial characteristics of cultural landscape genes. These include density, integration, and accessibility, which will enhance the scientific validity and generalizability of the results of the genetic characterization of cultural landscapes. Thirdly, the research will be expanded to Southeast China to enrich the cultural landscape gene pool and verify the applicability of landscape gene theory in different cultural contexts.

Through a combination of quantitative and qualitative methods, this research details the specific cultural landscape genes of Linpu Village. It demonstrates how these genes can inform effective conservation and development strategies. This approach not only promises sustainable solutions for the preservation and enhancement of traditional villages globally, but also provides a scalable model that showcases the broader applicability of cultural landscape gene theory in cultural landscape preservation.

2. Literature Review

To further deepen the understanding of cultural landscape gene theory, this section thoroughly explores the research overview of settlement landscapes, the basic concepts of cultural landscape gene theory, its historical development, and its application in modern research. This comprehensive examination aims to better grasp the practical significance and potential application of cultural landscape gene theory in global village conservation strategies.

2.1. Overview of Research on Settlement Landscapes

The study of settlement landscapes can be traced back to the 19th century, with Johann Georg Kohl's systematic study from a geographical perspective, which made settlement landscapes a hotspot in academic research [14]. By the 1980s, the purpose of conservation shifted from purely historical significance to heritage culture, and the study content began to diversify gradually [15]. Entering the 21st century, interdisciplinary convergence became increasingly evident, concentrating research on several major fields, including the economic promotion of settlement development [16], spatial research on settlement landscapes [17], and the sustainable development of settlements [18]. Recently, international research has focused on settlement landscape characteristics [19], the ecological construction of settlements [20], and the spatial patterns of settlements [21,22]. These studies have promoted the development of cultural diversity in settlement landscapes and provided a rich practical basis for applying cultural landscape gene theory.

2.2. Underlying Concepts

To better understand the theory of cultural landscape gene, it is necessary to clarify a few relevant foundational concepts (Table 1).

Table 1. Comparison of Genes, Cultural Genes, CLGTS, and Landscape Genetics.

Comparison Item	Genes	Cultural Genes	CLGTS	Landscape Genetics
Concept Origin	Biology, Genetics	Genetic Theory, Cultural Heritage	Cultural Gene Theory, Human Geography	Landscape Ecology, Population Genetics
Definition	Hereditary unit controlling biological traits, with replication, mutation, and variation	The basic unit of cultural transmission, with replication, variation, and selection	Cultural elements in traditional settlements, with uniqueness, replication, and variation	Study of how landscape features affect gene flow, population structure, and species adaptation
Application Fields	Biology, Medicine, Agriculture, Evolutionary Studies	Social Sciences, Cultural Studies, Information Dissemination	Traditional Settlement Preservation, Rural Planning, Cultural Heritage Protection	Biodiversity Conservation, Species Management, Ecosystem Restoration
Research Methods	Laboratory Experiments, Gene Sequencing, Molecular Biology Techniques	Observation, Experiments, Theoretical Modeling, Interdisciplinary Methods	GIS, Remote Sensing, Field Surveys	GIS, Remote Sensing, Genomic Tools, Mathematical Modeling
Research Background	Understanding genetic mechanisms and biological evolution, studying genetic diseases, and breeding	Exploring mechanisms of cultural transmission and evolution, understanding cultural and information dissemination patterns	The need for cultural heritage protection in the face of rapid urbanization	The impact of landscape changes on the genetic structure and connectivity of natural populations
Main Goals	Understanding genetic mechanisms, improving breeds, treating genetic diseases	Explaining cultural phenomena and their dissemination and evolution, understanding information and cultural dissemination patterns	Extracting and protecting cultural landscape genes in traditional villages to achieve sustainable cultural heritage development and revival	Understanding and managing the impact of landscape features on gene flow and species adaptation

Genes: The concept of a "gene" originated in the 1960s, when Gregor Mendel proposed that hereditary factors are the key units controlling organism traits. The term "gene" officially appeared in the 1909 book *Principles of Precision Genetics* by Danish geneticist Wilhelm Johannsen [23].

Cultural Genes: Inspired by the hereditary properties of genes, social scientists began exploring the basic elements of cultural inheritance and their properties. In the 1950s, Alfred L. Kroeber and Clyde Kluckhohn hypothesized "cultural genes", arguing that they are the basic units of human culture, subject to self-replication, "mutation", and variation [24]. Richard Dawkins's 1976 book *The Selfish Gene* introduced the concept of "meme", which brought the idea of cultural genes into social science research, later leading to the theory of "memetics" [25]. Currently, the main research areas related to modality theory include linguistic and network modality research [26], providing valuable insights for the genetic study of cultural landscapes [11].

Landscape Genetics: This interdisciplinary field combines landscape ecology and population genetics to understand how landscape features affect gene flow, population structure, and species adaptation [27]. Conceptualized by Manel et al. and formalized as a research area, landscape genetics has made significant progress utilizing GIS, remote sensing, and genomics tools [28]. Despite the similarity in name, landscape genetics and cultural landscape genes focus on distinctly different areas. The former studies natural populations and their genetic connectivity influenced by landscape structure [29], while the latter explores the cultural transmission and evolution of traditional human settlements and cultural landscapes, examining culture–environment interactions in human contexts [30].

2.3. Formation and Development of Cultural Landscape Gene Theory

The concept of the cultural landscape gene is mainly inspired by cultural genes. Chinese scholar Liu Peilin combined the concept of "cultural genes" with practical research in human geography and related disciplines, proposing the concept of "cultural landscape genes". He posits that cultural landscape genes distinguish one cultural landscape from another, akin to biological genes' uniqueness, replicability, and variability [31]. Strictly speaking, landscape genes refer to cultural factors in traditional settlements with specific material carriers or specialized reference objects [11].

From a methodological perspective, the cultural landscape gene analyzes the spatial imagery characteristics of traditional settlement landscapes and their geographic significance, drawing on cultural gene theory. Historically, research results before 2015 were relatively few, focusing mainly on landscape gene characteristics and concentrating on Liu Peilin and his academic team [11]. Post 2015, research has spread to various institutions and individuals, expanding the thematic scope. Currently, landscape gene research spans three aspects: architectural culture [32], settlement patterns [33], and regional culture [34], primarily focusing on traditional settlements with distinctive landscapes across China, particularly in the southwest and central regions [11]. Research scales include architectural monoliths, architectural groups, villages, and settlements [12,35]. The research content is divided into three major fields: cultural landscape gene identification and classification, map construction and expression, and the practical application of landscape genes.

Cultural Landscape Gene Identification and Classification: This field extracts the most characteristic landscape genes by deconstructing landscape elements of traditional settlements. Liu Peilin pioneered four basic principles for identifying a cultural landscape gene: "intrinsic uniqueness, extrinsic uniqueness, local uniqueness, and overall dominance" [36]. He elucidated the biologically similar characteristics of traditional settlement cultural factors, significantly impacting subsequent research. Shen Xiuying et al. proposed four extraction methods—pattern, element, structure, and meaning—based on these principles [37]. However, this method has operational limitations, particularly for extracting unique landscape elements and cultural factors. Hu and colleagues later proposed a feature deconstruction–extraction method, distinguishing cultural landscape elements by category, merging similar elements, and classifying them into four categories of feature genes, optimizing extraction and classification methods [38].

Mapping Construction and Expression: Cultural landscape gene mapping involves researchers processing the identified and extracted cultural landscape gene information of traditional settlements, coding and recording it, importing it into a basic database, and visualizing it with GIS tools [30]. Cultural landscape gene mapping contains information such as spatial positioning, carrier history, and culture [33]. Currently, most research on the cultural landscape genes of traditional settlements stops at the construction stage, with fewer studies combining quantitative analysis methods to reveal spatial characteristics of landscape genes [39].

Practical Application of Cultural Landscape Genes: Application mainly focuses on traditional settlement landscape inheritance and protection and minority settlement cultural landscape gene analysis [40]. For instance, Liu Peilin remodeled the image of village cultural landscapes based on extracting and identifying cultural landscape genes, presenting each attraction as a complete thematic tour line to shape a recognizable cultural tourist place [41]. However, most studies are limited to extracting and identifying cultural landscape genes and collecting and integrating local cultural and geographic information data, without further application to traditional cultural inheritance and revitalization or proposing specific, effective paths and methods compared to past efforts [39].

To summarize, the theory of landscape genetics has moved from initial exploration to rapid development. However, several issues still need scrutiny and resolution:

Improvement of Analytical Methods: Current research mainly employs qualitative analysis through literature reviews and field research to confirm the characteristics of cultural landscape genes [11]. Further analysis of spatial characteristics is limited [10].

Introducing mathematical models and spatial analysis techniques is necessary to deeply analyze spatial distribution laws and dynamic changes, providing a scientific basis for protecting and revitalizing cultural landscape genes.

Deepening Practical Applications: Most research stops at collecting raw data on cultural landscape genes of traditional villages [42], lacking depth in processing and utilizing these data for further planning and application in sustainable village development [43]. Future research should focus on specific applications of cultural landscape genes in ecological protection, cultural tourism, and community development, proposing practical operational paths and methods to transform theoretical results into practice and achieve sustainable rural development.

Addressing Research Limitations and Geographic Imbalance: Current research primarily focuses on China, particularly traditional villages in the southwest and central regions [44,45]. This concentration limits the application and generalizability of the cultural landscape gene theory, hindering the summarization of revitalization paths for traditional villages globally. Expanding the scope to cover diverse cultural landscapes and traditional villages worldwide is crucial for building a comprehensive and universal theoretical framework for cultural landscape genes. Additionally, international cooperation should be strengthened to promote the application and development of cultural landscape gene theory on a global scale, exploring paths for protecting and revitalizing traditional villages in other countries and regions.

3. Methods

3.1. Study Region

Linpu Village, situated in Chengmen Town, Cangshan District, Fuzhou City, Fujian Province, is in the northeast corner of Nantai Island. It is bordered by the Lian River to the north and Jiuqu Mountain to the south, with a view of Gushan Mountain across the river. This geographic setup forms a scenic landscape where mountains and water intertwine seamlessly, as illustrated in Figure 1. With a dense population of approximately 7000 across its 2.9 km^2, the village comprises four administrative subdivisions: Lianjiang, Shishan, Fulian, and Shaoqi [46].

Historically, Linpu has been a vibrant community since the Ming Dynasty, with a notable legacy of contributing scholars to imperial courts. By the Qing Dynasty, leveraging its advantageous location for land and water transport, it evolved into a critical hub within the Minjiang River Basin for waterborne trade, agriculture, and the broader socioeconomic development of the region. This development is documented in various historical records. Linpu Village serves as a repository for diverse regional cultures, encompassing the Shangshu culture and the patriotic culture of the Southern Song Dynasty, among others. This blend aligns with the traditional village ethos prevalent in Fuzhou and highlights Linpu's distinct cultural identity. It boasts 14 cultural heritage units and 32 listed protected buildings. Recognizing its historical and cultural significance, Linpu Village was designated as a historical and cultural village of note in Fujian Province in 2007. It was included in the seventh batch of national-level historical and cultural villages in 2019 [47]. This distinction places it among the few millennium-old villages surviving within an urban context in China.

However, recent rapid urbanization in Fuzhou has posed significant threats to Linpu's architectural integrity and landscape. Unplanned development and insufficient regulatory oversight have led to substantial alterations in village texture and aesthetics, characterized by intrusive construction and the encroachment of new development zones. These changes urgently call for concerted efforts to preserve the traditional landscapes that embody Linpu's historical and cultural essence.

(**a**) Location of Fujian Province in China

(**b**) Location of Linpu Village in Fujian Province

(**c**) Spatial layout plan of Linpu Village

(a)The map available at http://bzdt.ch.mnr.gov.cn/, which is publicly accessible and not subject to copyright restrictions.
(b,c) This map was generated using ArcGIS software, version 10.6. For more details, please visit https://www.esri.com/en-us/arcgis/about-arcgis/overview.

Figure 1. Location and overall landscape of Linpu Village.

3.2. *Research Methods*

Identifying and extracting cultural landscape genes involve extracting elements or factors that reflect regional cultural features from complex landscape phenomena. Due to this, any element or factor can be considered a cultural landscape gene theoretically; the identification and extraction of cultural landscape genes must follow certain principles and theoretical methods. Liu Peilin introduces four fundamental principles for extracting cultural landscape genes from his research on traditional Chinese settlement landscapes, namely "intrinsic uniqueness, extrinsic uniqueness, local uniqueness, and overall dominance" [36]. Moreover, he maintains that extracting cultural landscape genes from traditional village landscapes requires specific technical methods. The methods proposed by Hu Zui et al. [38] for extracting material and non-material cultural landscape genes based on the cultural landscape of Xuanzhou Ancient Town, as well as the methods proposed by Yin Zhiyi et al. [33] for Huangpi Dayu Bay, offer certain valuable guidance for this study. Based on the above identification principles and extraction methods, a cultural landscape gene extraction system for Linpu Village has been established to extract cultural landscape genes. After the extraction of cultural landscape genes, the principles of typology are adopted to classify and explain landscape elements, deriving corresponding cultural landscape genes. The understanding of village cultural landscape genes is further enhanced by constructing a landscape gene information chain. This spatial organizational form reflects certain cultural symbolic features of cultural landscapes [26]. For a more intuitive and accurate analysis of the spatial features of village cultural landscape genes, quantitative research methods such as spatial syntax theory and kernel density analysis were combined to fully explore the features of traditional villages and propose relevant strategies for the preservation and sustainable development of traditional village features. In line with the principles mentioned above, as well as theories and methods, the integrated research process of "extraction-construction-analysis-preservation" of cultural landscape genes was established, as shown in Figure 2.

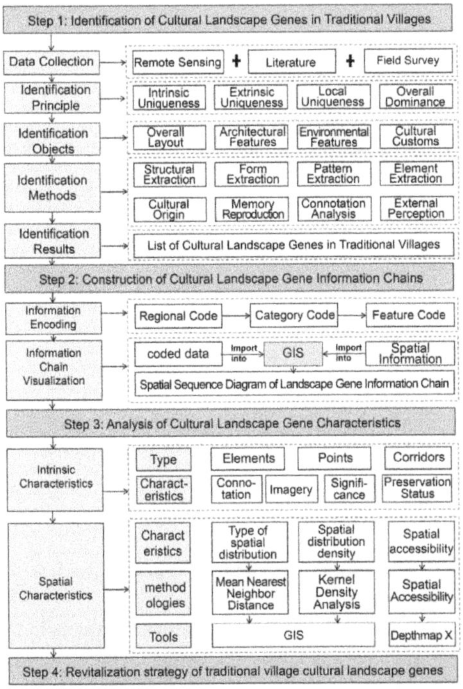

Figure 2. Research framework of sustainable inheritance of landscape features.

3.2.1. Identification and Encoding Methods of Cultural Landscape Genes

- Gene Identification and Extraction Method

The cultural landscape genes of Linpu Village mainly fall into two categories based on different material forms: material cultural genes and non-material cultural genes. The methods for extracting material elements include structural extraction, form extraction, pattern extraction, and element extraction. In contrast, those non-material elements include cultural context tracing, memory reproduction, connotation interpretation, and external perception [48]. The specific means of identification of cultural landscape genes is mainly through the visual and perceptual analyses of the researchers, specifically the perceptual identification of the cultural landscape genes of the villages through a variety of ways, such as viewing, field visits, surveys, reviewing historical and cultural data, and examining the characteristics of the natural environment. Based on the principles of typology and the "Guidelines for Strengthening the Preservation of Chinese Traditional Villages" [49], the identification and extraction results of material cultural landscape genes are classified into three identification factors, including overall layout, architectural features, and environmental references, and 10 identification indicators, including village form, street layout, and architectural style, etc. In comparison, the non-material cultural landscape genes are formed based on field research and classified into three identification factors, including regional culture, custom activities and traditional crafts, and four identification indicators, which ultimately establish the identification system for cultural landscape genes in Linpu Village (Table 2).

Table 2. Cultural landscape gene identification system in Linpu Village.

Gene Category	Identification Factor	Identification Sub-Indicators
Material cultural genes	Overall Layout	Landscape Pattern
		Street Layout
		Village Form
	Architectural Features	Plane Layout
		Public Building
		Architectural Style
		Architectural Decoration
		Architectural Texture
	Environmental References	Cultural Environment
		Natural Environment
Non-material cultural genes	Regional Culture	Surname Clan
		Object of Faith
	Custom Activities	Entertainment Customs
	Traditional Crafts	Inheritance Level

- Gene Information Chain Construction Method

This method is used to accurately categorize and elucidate the affinities among various cultural landscape genes, as well as to systematically record and manage the corresponding data. Based on the identification results of cultural landscape genes in Linpu Village, the encoding structure of cultural landscape genes is divided into region, category, and feature codes. The region code for Linpu Village is set as X. The category code, according to Richard Dawkins's concept of "Meme" [50], is set as M to represent cultural landscape genes, while distinguishing material cultural genes as M1 and non-material cultural genes as M2, respectively. The feature code follows the "Classification and Code of Fundamental Geographic Information feature" [51], categorizing the cultural landscape gene information of Linpu Village into "primary elements, secondary elements, tertiary elements, and landscape elements" and encoding them in Arabic numerals [36] (Figure 3). Our research team, which includes experienced researchers in relevant fields, conducted the cultural landscape gene encoding for Linpu Village. Our encoding method is based on established techniques referenced in several vital studies [33,52], which have been applied by other scholars in similar contexts. To ensure the accuracy and consistency of our encoding, two team members independently conducted the encoding process. Their results were compared, and discrepancies were discussed and resolved through consensus. Senior researchers reviewed and finalized this step to ensure alignment with established standards.

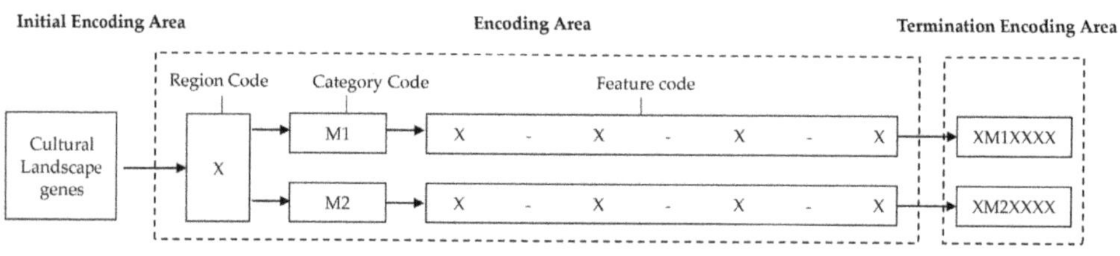

Figure 3. Encoding structure of cultural landscape genes.

3.2.2. Analysis Methods for Spatial Features of Cultural Landscape Genes

This study employs multiple analysis methods to comprehensively understand the spatial features of cultural landscape genes in Linpu Village. These methods include analyzing the spatial distribution types, density, and accessibility of cultural landscape genes. Through these methods, we can reveal the spatial organization and distribution patterns of cultural landscape genes, assess the concentration of their cultural value, and optimize their accessibility, thereby formulating scientific conservation and sustainable development strategies. These analyses also help identify areas that need prioritized protection and provide data support for improving infrastructure and enhancing the visitor experience.

- Analysis Method for Spatial Distribution Types of Genes

To evaluate the spatial organization of cultural landscape genes, this study first needs to determine their distribution types. Understanding the spatial arrangement of cultural landscape genes within the village can reveal the influence of natural and human factors on their distribution, aiding in developing targeted management and conservation strategies. Therefore, we employed the "average nearest neighbor" (*ANN*) method. On a macro level, the cultural landscape gene units in Linpu Village are point elements. The spatial distribution types of point elements can be categorized into three types: random, clustered, and uniform. The "average nearest neighbor" method is often used to determine the spatial distribution types of point elements [53], which is also suitable for determining the spatial distribution types of cultural landscape genes. The calculation formula for the average nearest neighbor (*ANN*) ratio is as follows:

$$ANN = \frac{\overline{D_O}}{D_E} \tag{1}$$

$$D_E = \frac{1}{\sqrt[2]{n/A}} \tag{2}$$

where $\overline{D_o}$ is the actual average distance between each cultural landscape gene unit and its nearest neighbor; D_E is the expected average distance between the cultural landscape gene units, assuming a random distribution pattern; n is the number of landscape gene units; and A is the village scope, including all cultural landscape gene units. When $ANN = 1$, the cultural landscape genes tend to be randomly distributed; when $ANN > 1$, the cultural landscape genes tend to be discrete; when $ANN < 1$, the cultural landscape genes tend to be clustered.

- Analysis Method for Spatial Distribution Density of Genes

To identify and prioritize areas with high cultural and historical value, this study needs to understand the spatial density distribution of cultural landscape genes. High-density areas typically indicate a concentrated distribution of cultural heritage, which is crucial for formulating graded conservation strategies and optimizing resource allocation. Therefore, we employed the kernel density analysis method [54]. The kernel density analysis method is commonly used to calculate the spatial density of point elements [53], and this method is also applicable for analyzing the spatial distribution density of cultural landscape genes. When calculating kernel density, cultural landscape gene units are assigned values according to the number of cultural landscape genes represented by the cultural landscape gene unit. Where a cultural landscape gene unit represents more than one cultural landscape gene, the landscape unit is weighted with a weight determined by the ratio of the number of cultural landscape genes in the unit to the total number of genes in all units, ensuring that units with a higher number of genes have a greater impact on the overall density estimate. This weighting approach helps emphasize culturally and historically significant areas due to a higher concentration of cultural landscape genes. The formula for calculating the kernel density, taking into account the weights (*W_KD*), is given below: The calculation formula for kernel density, considering the weights (*W_KD*), is as follows:

$$W\_KD(x) = \frac{1}{\sum_{i=1}^{n} W_i} \sum_{i=1}^{n} W_i K(\frac{x - X_i}{h}) \tag{3}$$

where x represents a position in the spatial domain for which we estimate the density. W_i defines the weight for the ith unit, calculated as the proportion of cultural landscape genes within that unit relative to the sum of cultural landscape genes across all units. K denotes the kernel function. X_i is the location of the ith cultural landscape gene unit. h stands for the bandwidth.

A spatial distribution model of cultural landscape genes in Linpu Village is constructed using ArcGIS version 10.6 (Esri, Redlands, CA, USA). The cultural landscape genetic elements within the village are assigned values, and the kernel density analysis method is adopted to calculate the spatial model of cultural landscape genes. The results of the kernel density calculation are presented in the form of images, and the gene density levels are graded based on the distribution results of landscape units. The color depth represents the difference in kernel density values, with darker colors indicating higher kernel density values.

- Analysis Method for Spatial Accessibility of Genes

Spatial accessibility analysis aims to examine the capacity of cultural landscape genes to attract traffic flow. High accessibility areas are more likely to attract visitors, promoting the display and conservation of cultural heritage. In traditional villages, cultural landscape genes are interconnected via existing traffic paths, constituting a transportation network. Within this network, the transportation cost associated with accessing each cultural landscape gene unit varies. The concept of "topological depth" quantifies transportation costs, whereas a higher topological depth value indicates increased costs for a cultural landscape gene unit and reduced accessibility. To quantify the spatial accessibility of cultural landscape gene units, the method of calculating their global integration was adopted. Global integration is the aggregate of the reciprocals of topological depth values from a given cultural landscape gene unit to all others within its historical and cultural village town. Consequently, spaces exhibiting high integration are characterized by enhanced accessibility [29]. This analysis was conducted using Depthmap X version 0.8.0 (Space Syntax Laboratory, The Bartlett School of Architecture, University College London, UK) to assess the global integration of Linpu Village's spatial axes. The formula to calculate the global integration of cultural landscape genes is as follows:

$$I_i = \frac{(n-2)D_n}{2\left(nDi^{-1}\right)} \tag{4}$$

where MD_i is the average topological depth value of cultural landscape gene i, and its calculation formula is:

$$MD_i = \frac{TD_i}{n-1} = \frac{\sum_{j-1}^{n} \mathrm{d}_{ij}}{n-1} \tag{5}$$

$$Dn = \frac{2\{n[log((n+2)/3)-1]+1\}}{(n-1)(n-2)} \tag{6}$$

n is the total number of cultural landscape genes in the historical and cultural village. D_N is the mean distance between cultural landscape genes; TD_i is the total depth of the ith cultural landscape genes, summing the steps from that gene to all others; and d_{ij} is the distance from the ith cultural landscape gene to the jth cultural landscape gene.

3.3. Data Sources

The data required for this study and their sources include:

(1) Remote Sensing Image Data

The 2020 remote sensing images of Linpu Village, obtained from ResourceSat-1, were sourced from the Geospatial Data Cloud site [55]. These images were collected using the vectorization of geographic information, including buildings, road networks, and rivers, using ArcGIS version 10.6. This process aids in analyzing and interpreting the village's natural environment and spatial configuration. These data provide the foundation for the subsequent identification of cultural landscape gene locations in Linpu Village and the visualization of the quantitative analysis of cultural landscape gene spatial features.

(2) Literature Data

Various literature sources were collected, including documents such as the "Historical and Cultural City Preservation Plan of Fuzhou City" [56], "Historical and Cultural City, Town, and Village Chronicles of Fuzhou City" [57] and "Preservation Plan for Linpu Village in Chengmen Town, Cangshan District, Fuzhou City, a Historical and Cultural Village of Fujian Province" [58], to analyze the cultural features of Linpu Village. The data obtained from the literature review were organized and summarized using the literature analysis method.

(3) Field Surveys

The field surveys comprised two main components: observational surveys and structured interviews. These were conducted to gather comprehensive data on the cultural landscape genes and the villagers' perception of their cultural heritage.

Survey Process: Field surveys were conducted between January and February 2023 to observe the physical and cultural landscape features of Linpu Village directly. These involved on-site visits to various cultural landscape gene locations to document their current state, spatial characteristics, and accessibility. GPS devices and cameras were used to record precise locations and visual conditions.

Interviews: Semi-structured interviews were conducted with 36 participants, including 27 villagers, 2 village committee members, and 7 tourists. Each interview lasted between 10 and 90 min, depending on the depth of the discussion and the availability of the interviewee. Villagers were asked about their views on the protection of traditional culture, opinions on tourism development, perceptions of changes in the village landscape, and challenges in maintaining cultural heritage. Village committee members were asked about the basic conditions of the village, the maintenance of cultural heritage, and the integration of cultural preservation with development plans. Tourists were asked about their sensory experiences, perceptions of the village's cultural landscape, and suggestions for improving the visitor experience (Table 3). The interviews aimed to understand the significance of various landscape genes from the perspective of the community, capturing their views on preservation needs and future development. These insights were then used to inform the development of strategies for revitalizing the cultural landscape genes, ensuring they meet the needs and expectations of the village.

Table 3. Interviewee information and focus of interviews.

Code	Interviewee Group	Number of Participants	Description	Key Focus of Interviews
RI~R27	Villagers of Linpu Village	27	R1~R13 (ordinary villagers), R14~R19 (shop owners), R20~R23 (stallholders), R24~R25 (intangible cultural heritage bearers), R25~R27 (guesthouse receptionists)	Views on the Protection of Traditional Culture and Tourism Development of Villages
G1~G2	Village Committee Members	2	G1 (Village Party Secretary), G2 (Women's Director)	Basic conditions of the village and maintenance of culture
VI~V7	Tourists	7	V1~V7 (Tourists)	Sensory experiences of tourists

Note: 'R' denotes villagers, 'G' for government workers, and 'V' for tourists.

4. Results

4.1. Identification and Encoding of Cultural Landscape Genes

4.1.1. Identification and Extraction of Cultural Landscape Genes in Linpu Village

The identification results of cultural landscape genes are shown in Table 4. In terms of overall layout, Linpu Village exhibits a landscape pattern featuring mountains and water, a network layout of streets and alleys, and a clustered village form. In terms of architectural features, Linpu Village's buildings are mainly traditional buildings from the Ming Dynasty, Qing Dynasty, and the Republican period, with typical features of Fuzhou residential buildings. In terms of environmental references, the cultural heritage of Linpu Village is mainly reflected in cultural relics such as memorial archways, cliff inscriptions, and ancient bridges. At the same time, the natural environment is represented by ancient trees and water systems. Regarding intangible cultural genes, these mainly include regional cultures represented by Shangshu culture, hero worship, and religious beliefs, custom activities represented by welcoming the God of Mount Tai and the pageant of gods, and traditional crafts represented by Annan instrumental performance.

Table 4. Identification of cultural landscape gene in Linpu Village.

Category	Factor	Indicators	Result	
Material cultural genes	Overall layout	Landscape Pattern	Surrounded by mountains and waters	
		Street Layout	Grid structure	
		Village Form	Clustered	
	Architectural features	Plane Layout	Multiple-courtyard style	
		Public Buildings	Ancestral temple	
			Ancient academy	
			Clan hall	
		Architectural Style	Facade Form	Double-slope roof with dovetail ridge
			Gable Design	Corresponding to the five-element theory
			Roof Truss Structure	Column and tie construction
		Architectural Decoration	Decoration techniques	Wood carving, stone carving, brick carving, etc.
		Architectural Texture	Building materials	Rammed earth walls with gray tiles
	Architectural Features	Architectural Texture	Color	Mainly gray and brown
	Environmental references	Cultural Environment	Memorial archway	
			Ancient bridge	
			Cliff inscriptions	
		Natural Environment	The Lian Jiang River	
			Centuries-old banyan tree	
Non-material cultural genes	Regional culture	Surname Clan	Shangshu culture (genealogy culture)	
		Object of Faith	Hero worship	
			Religious belief	

Table 4. *Cont.*

Category	Factor	Indicators	Result
Non-material cultural genes	Custom activities	Entertainment Customs	Welcoming the God of Mount Tai
			Pageant of gods in Ruiji Temple
	Traditional craft	Level of Inheritance	Annan instrumental performance

4.1.2. Construction of Cultural Landscape Gene Information Chain in Linpu Village

The independent encoding results from the two team members showed a high degree of consistency. Initial comparisons revealed minor discrepancies in categorization, primarily due to subjective interpretations of certain landscape features. These discrepancies were thoroughly discussed in joint sessions, where the team members reconciled differences through consensus, ensuring a unified and accurate final encoding.

Figure 4 provides examples of the encoded cultural landscape genes, demonstrating the application of our method. These examples illustrate the robustness and applicability of our coding framework. The final encoding results reflect a comprehensive and consistent categorization of the cultural landscape genes in Linpu Village, as validated by senior researchers.

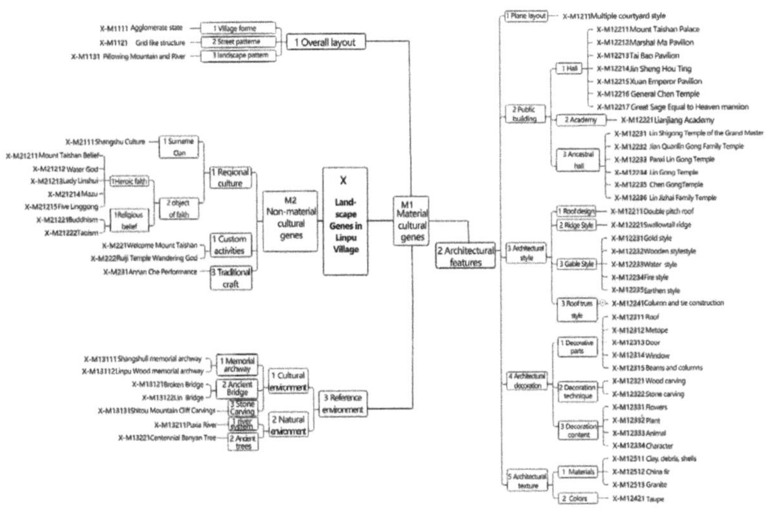

Figure 4. Cultural landscape gene information chain in Linpu Village.

On this basis, the spatial location of cultural landscape gene elements was extracted using a geographic information system (GIS), providing parameters such as ID name, coding information, coordinate information, etc.; the cultural landscape genes and the spatial distribution characteristics of cultural landscape genes were visualized (Figure 5). It can be drawn from the construction results that there are four landscape corridors in the cultural landscape gene information chain of Linpu Village, one running north to south and three running east to west. The grid-like layout of the streets and alleys reflect the overall layout gene, the architectural feature gene and regional culture gene are mainly distributed in the northern and central–southern parts of the village, and the reference environment gene is distributed primarily in the northern and southern parts of the village.

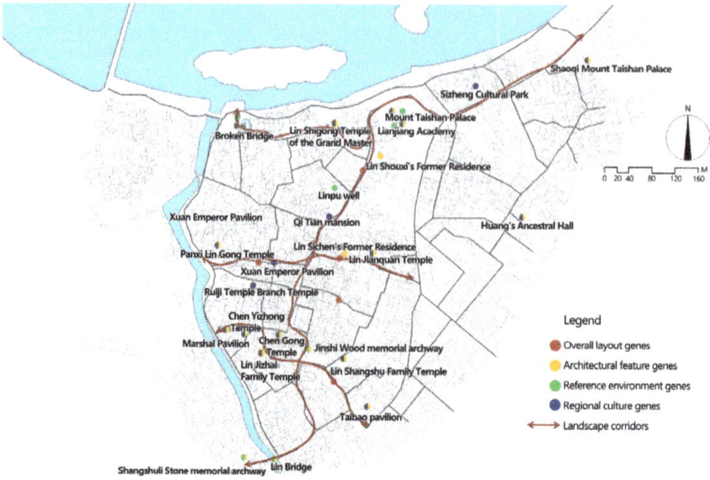

Figure 5. Spatial sequence diagram of cultural landscape gene information chain in Linpu Village.

4.2. *Analysis of Intrinsic Characteristics of Cultural Landscape Genes in Linpu Village*

The landscape information elements are the cultural elements of a landscape, while the landscape information points embody these elements. Landscape information elements and points serve as the fundamental elements and units for constructing a landscape information chain. In contrast, landscape information corridors represent the spatial manifestation of the landscape information chain [59], as shown in Figure 6. Through the identification and extraction of the cultural landscape genes in Linpu Village (Table 4), the main features of the information elements, information points, and information corridors in Linpu Village can be derived from the construction of the cultural landscape gene chain (Figure 4) and information graph (Figure 5).

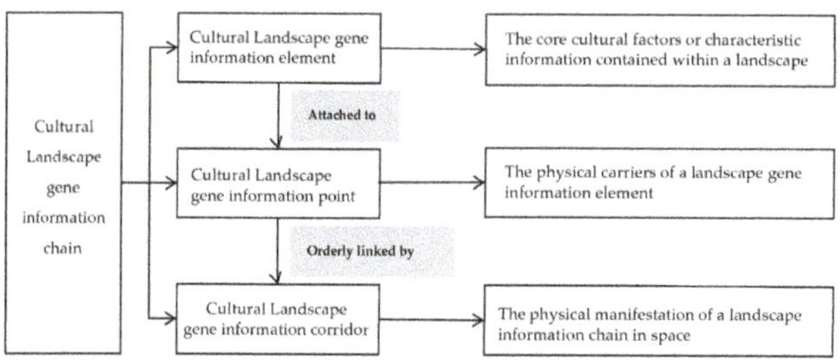

Figure 6. The intrinsic composition of cultural landscape gene information chain.

4.2.1. Features of Cultural Landscape Gene Information Elements

The landscape information elements in Linpu Village include the Shangshu culture, the patriotic culture of the Southern Song Dynasty, clan culture, religious culture, and folk culture. These elements are deeply rooted in the local cultural perspectives, enriching a rich historical context of the village's culture with the splendid "Shangshu Culture" of the Ming Dynasty and the patriotic culture of the Southern Song Dynasty. This amalgamation creates a complex historical context that shapes the village's cultural landscape. However, the village faces challenges in maintaining this heritage. Traditional customs such as clan sacrifices are in decline due to the loss of genealogies and fading clan consciousness. Similarly, the art of Annan instrumental performance is at risk as the older generation of performers dwindles and the younger generation shows less interest. Additionally, there is an imbalance in how cultural diversity is displayed. While there is a significant emphasis on Shangshu and patriotic cultures, lesser-known cultural elements and regional crafts receive limited attention, risking a loss of cultural diversity.

4.2.2. Features of Cultural Landscape Gene Information Points

The landscape information points in Linpu Village include architectural structures such as Taishan Temple and Lianjiang Academy that embody the features of Shangshu culture and patriotic culture, as well as residential houses and ancestral temples that showcase folk culture and clan culture, and architectural relics representing diverse cultures. The existing buildings mainly reflect the traditional courtyard architecture of Ming Dynasty, Qing Dynasty, and Republican periods, while the new constructions have an impact on the overall style and features [60]. The natural landscapes, such as water systems and ancient trees, blend with cultural landscapes such as cliff stone inscriptions, broken bridges, and Lin Bridge, bearing historical research value.

4.2.3. Features of Cultural Landscape Gene Information Corridors

The internal corridors in Linpu Village are divided into primary and secondary corridors. The primary corridor is Linpu Avenue, which runs from south to north, serving as the basic framework of the village's layout. It connects Yudao Street to the north and stretches from Jin Shenghou (Qianliyan) Pavilion to Shaoqi Ferry to string various nodes in the village together, such as the Shangshuli Memorial Archway and Qitian Temple along Yudao Street, ensuring a complete cultural connotation to Linpu Village. The secondary corridor is the road network that links the internal residential buildings, forming a cultural tourism belt of Linpu that connects residential buildings, the ancestral temple, and the clan temple. The main and secondary corridors intersect with ancestral temples, residential houses, and community temples, forming a "fishbone-shaped" street layout in Linpu Village [19]. Nowadays, the original green slab stone on the corridors of Linpu Village has been replaced by cement pavement. The modern-style buildings constructed on both sides of the corridors are incongruous with the entire village's traditional style, which has affected the consistency and coherence of the landscape information chain and undermined its integrity in Linpu Village. Compared with the primary corridor, the marginal landscape information corridor that is supported by the remaining alleys, such as Houjie Street and Taibao Lane, and connects the landscape information points, such as pavilions and the clan temple, bears a lower readability of cultural landscape information and underestimated value, leaving some potential points of landscape information unexplored.

4.3. Analysis of Spatial Features of Cultural Landscape Genes in Linpu Village

4.3.1. Analysis of Spatial Distribution Types of Cultural Landscape Genes

The "nearest neighbor ratio" method in ArcGIS is adopted to analyze the spatial distribution types of cultural landscape genes in Linpu Village. The calculation results show that the nearest neighbor ratio of cultural landscape genes in Linpu Village is 1.1815, which is greater than 1, indicating that the natural geographic conditions with uniform landscape elements in the village are the main factors affecting the spatial distribution types of cultural landscape genes. Linpu Village is located in a flat coastal delta region, with regular and symmetrical residential clusters, which also contributes to the uniform spatial distribution of cultural landscape genes. However, this dispersion, lack of centralization, and clear spatial layout make it difficult for tourists to fully and systematically grasp and understand the cultural characteristics of Linpu Village.

4.3.2. Analysis of Spatial Distribution Density of Cultural Landscape Gene

After thoroughly evaluating the spatial scale of landscape patterns and the distribution of cultural landscape gene units within our study area, we selected a bandwidth of 20 km for the kernel density estimation. This bandwidth was chosen to adequately capture the broad-scale spatial patterns of cultural significance without overemphasizing local variations that might result from isolated or outlier cultural landscape gene units. It strikes a balance between providing sufficient detail to discern meaningful clusters of cultural landscape genes and ensuring a smooth density surface that reflects the overarching distribution trends. This choice is consistent with the spatial extent of our study area and the average dispersion of cultural landscape gene units, thereby allowing us to generate a density map that accurately represents the density gradients of cultural heritage significance across the landscape. The calculation results of kernel density of cultural landscape genes indicate that the density of landscape information points in Linpu Village exhibits a clustered distribution, specifically characterized by the "one core, two clusters" pattern (Figure 7). The highest kernel density is located in the northern part of Linpu Village, forming the "gateway core area" composed of Taishan Temple and Lianjiang Academy, reaching the highest level of 9. The other two clusters are located in the central "residential and ancestral temple cluster" and the southern "diverse culture cluster", with cultural landscape gene densities ranging from levels 2 to 5. Apart from these concentrated areas, the distribution of cultural landscape genes in the surrounding village is more scattered, with densities mainly ranging from levels 1 to 3.

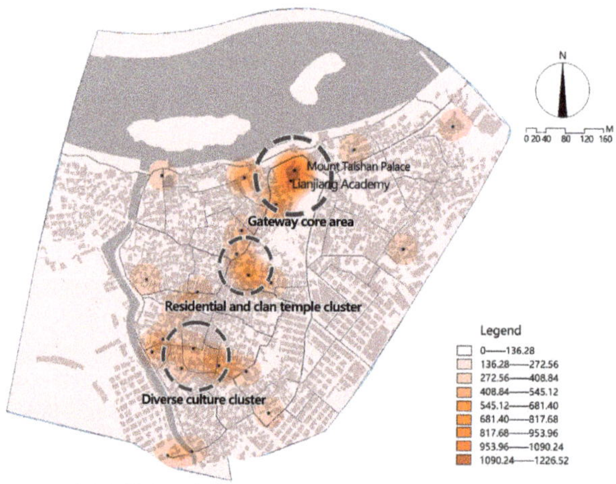

This map was generated using ArcGIS software, version 10.6. For more details, please visit https://www.esri.com/en-us/arcgis/about-arcgis/overview.

Figure 7. Kernel density map of cultural landscape genes in Linpu Village.

4.3.3. Analysis of Spatial Accessibility of Cultural Landscape Genes

The calculation results of the global integration of cultural landscape genes in Linpu Village show a gradual decrease in spatial accessibility from the center to the periphery (Figure 8). The areas with the highest global integration in Linpu Village are concentrated around Linpu Street, Jinshi Lane, Houjie Street, and their surroundings, corresponding to the highest spatial accessibility of cultural landscape genes near the "residential and ancestral temple cluster". The northern part, represented by Taishan Temple and Lianjiang Academy, is the "gateway core area" that embodies the Shangshu culture and the patriotic culture of Linpu Village, and, with the southern "diverse culture cluster", are the two areas with higher concentrations of cultural landscape genes in the village. Despite these features, both areas show comparatively low global integration, which limits visitor engagement with these culturally rich locations and adversely affects the overall portrayal of Linpu's landscape attributes. This disparity poses challenges for the effective communication and study of the village's cultural heritage.

This section may be divided by subheadings. It should provide a concise and precise description of the experimental results, their interpretation, and the experimental conclusions that can be drawn.

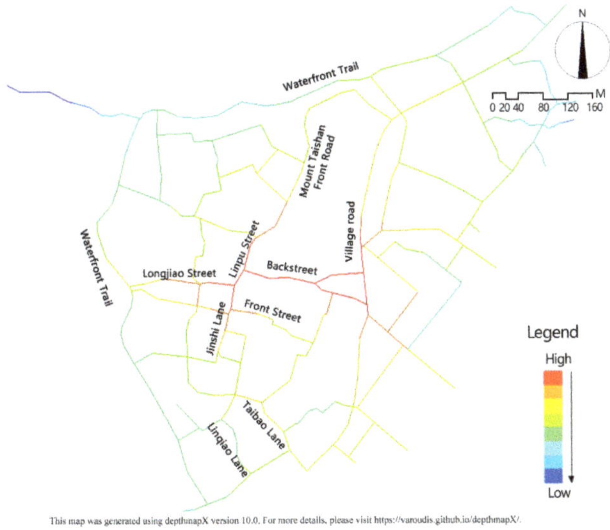

Figure 8. Global integration analysis of Linpu Village.

5. Discussion

This study systematically identified and analyzed the cultural landscape genes of Linpu Village, employing a method and coding system that have been adopted by other researchers [33,52], thus enhancing the reliability and scientific rigor of our approach. By combining an in-depth examination of the connotative characteristics of these genes with an innovative spatial attribute quantification method—including spatial distribution type, density, and accessibility—we address limitations present in previous studies that focused solely on connotative analysis [11].

Our findings reveal that Linpu Village's cultural landscape genes are diverse and rich, displaying typical Fuzhou characteristics and unique local charm. These attributes are crucial for the village's protection and revitalization. However, some genes risk disappearing, necessitating enhanced protection measures. Spatially, these genes exhibit a uniform group distribution with a pronounced core, and accessibility diminishes from the center to the periphery, hampering the effective dissemination and utilization of cultural information.

To facilitate the sustainable development of traditional villages, we propose integrating the connotative characteristics and spatial distribution patterns of cultural landscape genes and suggest revitalization strategies. The practical application of landscape gene theory remains in its infancy, with limited research available [39]. Existing studies primarily focus on connotative protection, with scant attention to strategies for revitalizing cultural inheritance [61]. Furthermore, there is a lack of empirical research on rural landscape planning and tourism development, necessitating deeper investigation [43].

Based on our comprehensive analysis, we propose a strategy to transform Linpu Village's cultural landscape resources into economic vitality. This strategy integrates cultural inheritance, spatial optimization, and industrial development to revitalize these resources sustainably. It aligns with Linpu Village's cultural distinctiveness, the well-being of the local population, and tourist expectations. The specific process is illustrated in Figure 9.

The integrated planning of connotative features and spatial distribution patterns is essential for the better protection and utilization of these resources. Using kernel density and accessibility analysis, we identified landscape nodes to ensure scientifically sound selection. Information corridors highlight Linpu Village's cultural axis, establishing a spatial pattern for protection. The central node, due to its high accessibility and rich cultural elements, should be the primary focus of development, equipped with public spaces, infrastructure, and landscape design. The northern node, reflecting Shangshu culture and patriotism, forms a corridor with riverside genes rich in natural and cultural attributes. The southern node embodies ancestral temple culture, forming a key cultural landscape. Future developments should restore and maintain these traditional characteristics. The "three horizontal and one vertical" corridors connect Linpu Village's cultural landscapes, providing tourists with a comprehensive traditional village experience. The preservation and enhancement of these corridors should be prioritized, as indicated in Figure 10.

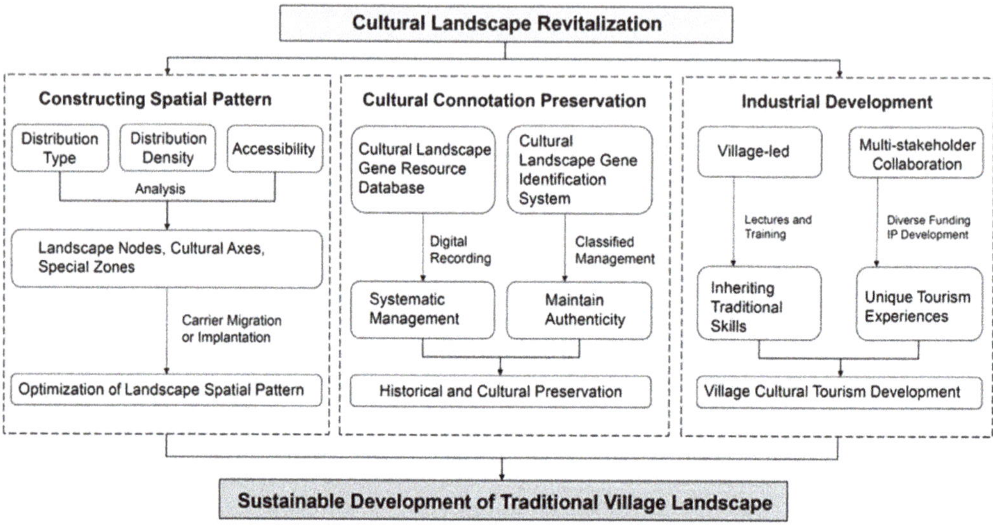

Figure 9. The model for the sustainable development of traditional village landscapes.

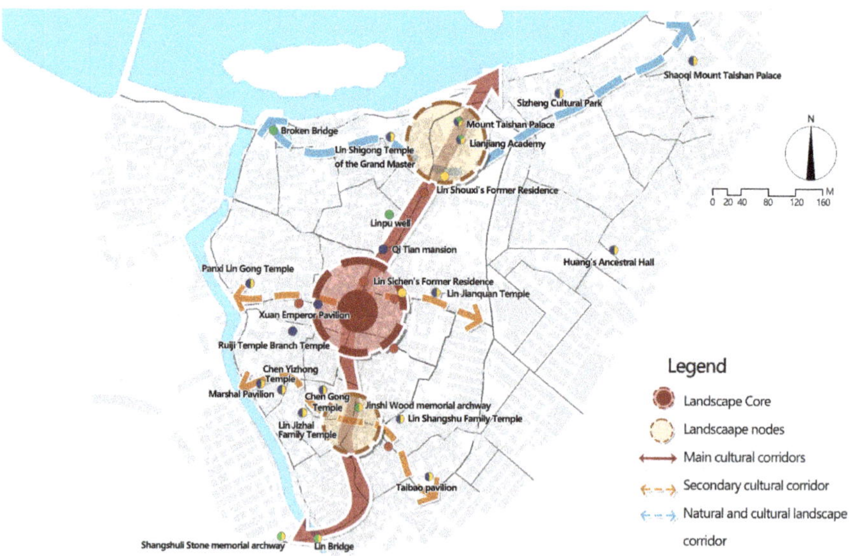

This map was generated using ArcGIS software, version 10.6. For more details, please visit https://www.esri.com/en-us/arcgis/about-arcgis/overview.

Figure 10. The pattern of the spatial protection of the village landscape.

Spatial optimization can be achieved through cultural landscape gene carrier migration and implantation. In regions with limited accessibility and low gene agglomeration, relocating cultural landscape gene carriers to higher concentration areas ensures better protection and resource utilization. Establishing historical and cultural corridors in accessible alley spaces and introducing new carriers can enhance traditional alley spaces and preserve the historical and cultural narrative, further strengthening the village's identity and sense of belonging.

For the protection of cultural connotations, a cultural landscape gene resource database should be established to systematically manage and digitally monitor Linpu Village's elements. By integrating multiple data sources, a 3D cultural landscape gene database can protect the village's history and culture's integrity, authenticity, and sustainability [42]. Establishing a cultural landscape gene identification system supplemented by manual diagnosis can examine the inheritable variation, clonal variation, or mutation of Linpu Village's cultural landscape genes [13]. Different conservation methods should be adopted based on the characteristics of these elements. Inherited cultural landscape genes refer to features and structures that have remained unchanged throughout their evolution, preserving original traits or historical continuity. Important inherited elements like Taishan Temple and Lianjiang Academy should maintain their integrity and continuity, preserving their original appearance and historical information. Clonal variations, which are morphological changes adapting to social or human-induced influences, need adjustment to align with the overall landscape. For instance, variations in traditional commercial streets' architectural forms and materials should be harmonized with surrounding landscapes. Mutated cultural landscape genes, which have transformed into completely new forms unrelated to their original state, such as modern restaurants and shops, should be replaced or refurbished to maintain Linpu Village's traditional cultural style.

Regarding tourism and cultural industry development, Linpu Village should learn from successful rural development experiences in Japan, emphasizing the central role of villagers [62,63]. Organizing cultural lectures and vocational training can transmit traditional skills and customs. Collaborating with businesses in tourism projects can diversify villagers' income. Leveraging Linpu Village's unique cultural landscape genes to create high-value cultural and economic products involving artists and designers can enhance visibility and attractiveness. Collaborating with digital media professionals can also build the village's cultural IP. Traditional tours should evolve into interactive dynamic models, creating thematic routes to enhance consistency and memorability. Regularly adding interactive platforms and cultural activities can encourage repeat visits and highlight the village's unique features.

This study's theoretical contributions extend beyond Linpu Village, offering significant insights into cultural landscape studies and rural development. The innovative methodological framework we developed, combining connotative and spatial analyses, can be applied to other rural contexts to assess and preserve cultural landscapes. Our approach integrates tangible and intangible heritage elements into a cohesive analytical model, enhancing the understanding of cultural landscapes' spatial dynamics. The strategies proposed for sustainable rural development provide a replicable model for other traditional villages, emphasizing the importance of cultural sustainability alongside economic and social dimensions. By validating our theoretical models in the practical context of Linpu Village, we bridge the gap between theory and practice, encouraging further application and refinement in diverse settings. This interdisciplinary research advances theoretical discourse across cultural studies, geography, spatial analysis, and rural development, promoting a holistic understanding of cultural landscapes and encouraging future interdisciplinary collaborations.

Despite the significant contributions of this study, several limitations need to be acknowledged. Firstly, the research is confined to a single village, Linpu, which may limit the generalizability of the findings. While the methodologies and frameworks developed are theoretically applicable to other contexts, further research is needed to test and validate these approaches in diverse cultural and geographical settings. Secondly, the data collection relied heavily on qualitative methods, which, while providing deep insights, may introduce subjectivity. Future studies could benefit from incorporating more quantitative data to enhance the robustness of the findings.

Additionally, the study primarily focuses on preserving and revitalizing cultural landscape genes without fully exploring the economic impacts of the proposed strategies. Longitudinal studies assessing the economic outcomes of implementing these strategies would provide valuable insights. Furthermore, the dynamic nature of cultural landscapes, influenced by ongoing social, economic, and environmental changes, necessitates continuous monitoring and adaptive management strategies. Future research should explore the development of real-time monitoring systems and adaptive management frameworks to ensure the sustainable preservation and utilization of cultural landscapes.

Finally, the role of community engagement and participation in the preservation and revitalization process was highlighted. However, further investigation is needed to understand the most effective ways to engage local communities and stakeholders in these efforts. Understanding the social dynamics and fostering strong community involvement is crucial for the success and sustainability of cultural landscape preservation initiatives.

By addressing these limitations and exploring these future research directions, we can further enhance our understanding and application of cultural landscape theory, contributing to the sustainable development and preservation of cultural heritage worldwide.

6. Conclusions

This study systematically identified and analyzed the cultural landscape genes of Linpu Village, utilizing a method and coding system widely adopted by other researchers, thereby enhancing the reliability and scientific rigor of our approach. By integrating connotative characteristics and spatial attribute quantification, our findings reveal the diverse and rich cultural landscape genes that exhibit typical Fuzhou characteristics and unique local charm. These attributes are crucial for the village's protection and revitalization.

Our research makes several key contributions to the field of cultural landscape studies and rural development. Firstly, the innovative methodological framework combining connotative and spatial analyses can be applied to other rural contexts, providing a comprehensive tool for cultural landscape research and planning. Secondly, the strategies proposed for transforming cultural landscape resources into economic vitality offer a replicable model for sustainable rural development, emphasizing the integration of cultural, economic, and social dimensions.

Despite these contributions, the study has several limitations. The focus on a single village, Linpu, may limit the generalizability of the findings. Future research should test and validate these approaches in diverse cultural and geographical settings. Additionally, the reliance on qualitative data introduces subjectivity, which can be mitigated by incorporating more quantitative data in future studies. Moreover, the economic impacts of the proposed strategies were not fully explored, suggesting the need for longitudinal studies to assess the long-term outcomes.

Future research should also develop real-time monitoring systems and adaptive management frameworks to address the dynamic nature of cultural landscapes. The effective engagement of local communities and stakeholders in the preservation and revitalization process remains a critical area for further investigation, as understanding social dynamics and fostering community involvement are essential for sustainable cultural landscape preservation.

In conclusion, this study advances the theoretical and practical understanding of cultural landscape genes and their role in sustainable rural development. By addressing the identified limitations and exploring the suggested future research directions, we can further enhance the application of cultural landscape theory, contributing to the preservation and development of cultural heritage worldwide.

Author Contributions: Conceptualization, S.L. and W.D.; Methodology, S.L. and L.W.; Software, L.W.; Validation, S.L. and C.X.; Formal analysis, L.W. and C.X.; Investigation, S.L., L.W. and W.D.; Resources, W.D.; Data curation, W.D.; Writing—original draft, L.W.; Writing—review and editing, S.L., L.W., C.X. and W.D.; Visualization, L.W. and C.X.; Supervision, W.D.; Project administration, S.L.; Funding acquisition, S.L. All authors have read and agreed to the published version of the manuscript.

Funding: This research was funded by (1) National Natural Science Foundation of China (NSFC), Grant No. 51778145; (2) National Natural Science Foundation of China (NSFC), Grant No. 52308054; (3) Fujian Provincial Social Science Foundation, Grant No. FJ2024B214; (4) Fujian Provincial Social Science Foundation, Grant No. FJ2023B100.

Data Availability Statement: Data are contained within the article.

Conflicts of Interest: The authors declare no conflict of interest.

References

1. Qiu, B.; Li, H.; Tang, Z.; Chen, C.; Berry, J. How cropland losses shaped by unbalanced urbanization process? *Land Use Policy* **2020**, *96*, 104715. [CrossRef]
2. Antrop, M. Landscape change and the urbanization process in Europe. *Landsc. Urban Plan.* **2004**, *67*, 9–26. [CrossRef]
3. Park, E.; Choi, B.-K.; Lee, T.J. The role and dimensions of authenticity in heritage tourism. *Tour. Manag.* **2019**, *74*, 99–109. [CrossRef]
4. Mokoena, L.G. Cultural tourism: Cultural presentation at the Basotho cultural village, Free State, South Africa. *J. Tour. Cult. Chang.* **2020**, *18*, 470–490. [CrossRef]
5. Thinh, N.K.; Gao, Y. Understanding the Informal Morphology of Villages-in-the-City: A Case Study in Hanoi City, Vietnam. *Sustainability* **2021**, *13*, 13136. [CrossRef]
6. Wang, T.; He, Q.; Wang, D.; Adeyeye, K.; Yu, P. Extension Theory for the Reconstruction of Traditional Villages: Case example in Dawa Village. In Proceedings of the 7th International Conference on Information Technology and Quantitative Management (ITQM)—Information Technology and Quantitative Management Based on Artificial Intelligence, Granada, Spain, 3–6 November 2019; pp. 191–198.
7. Ryberg-Webster, S. Heritage amid an urban crisis: Historic preservation in Cleveland, Ohio's Slavic Village neighborhood. *Cities* **2016**, *58*, 10–25. [CrossRef]
8. Jiang, Y.J.; Li, N.; Wang, Z.Y. Parametric Reconstruction of Traditional Village Morphology Based on the Space Gene Perspective-The Case Study of Xiaoxi Village in Western Hunan, China. *Sustainability* **2023**, *15*, 2088. [CrossRef]
9. Hu, Z.; Josef, S.; Min, Q.W.; Tan, M.; Cheng, F.L. Visualizing the cultural landscape gene of traditional settlements in China: A semiotic perspective. *Herit. Sci.* **2021**, *9*, 115. [CrossRef]
10. Hu, Z.; Wu, X. An Object-oriented Algorithm for Constructing the Conceptual Lattice of Cultural Landscape Genes of Traditional Settlements. *J. Geo-Inf. Sci.* **2024**, *26*, 604–619.
11. Jiang, S.; Fan, Y.; Zheng, W. Research Progress on Landscape's Gene Theory and Its Utilization in China. *J. Hum. Settl. West China* **2021**, *36*, 84–91.
12. Zheng, W.; Li, B.; Liu, P.; Zeng, R.; Deng, Y.; Zeng, C. Gene Identification and Zoning of Traditional Village Landscape Groups in Hunan Province. *Econ. Geogr.* **2021**, *41*, 204–212.
13. Xiang, Y.; Cao, M.; Qin, J.; Wu, C. Study of traditional rural settlements landscape genetic variability in Shaanxi Province based on accurate-restoration. *Prog. Geogr.* **2020**, *39*, 1544–1556. [CrossRef]
14. Muir, R. Geography and the history of landscape: Half a century of development as recorded in The Geographical Journal. *Geogr. J.* **1998**, *164*, 148–154. [CrossRef]
15. Zuo, D.; Li, C.; Lin, M.; Chen, P.; Kong, X. Tourism, Residents Agent Practice and Traditional Residential Landscapes at a Cultural Heritage Site: The Case Study of Hongcun Village, China. *Sustainability* **2022**, *14*, 4423. [CrossRef]
16. Haller, A.; Borsdorf, A. Huancayo Metropolitano. *Cities* **2013**, *31*, 553–562. [CrossRef]
17. Duncan, D.H.; Kyle, G.; Race, D. Combining facilitated dialogue and spatial data analysis to compile landscape history. *Environ. Conserv.* **2010**, *37*, 432–441. [CrossRef]
18. Owen, S.; Herlin, I.S. A Sustainable Development Framework for a Landscape of Dispersed Historic Settlement. *Landsc. Res.* **2009**, *34*, 33–54. [CrossRef]
19. Bazan, G.; Castrorao Barba, A.; Rotolo, A.; Marino, P. Vegetation series as a marker of interactions between rural settlements and landscape: New insights from the archaeological record in Western Sicily. *Landsc. Res.* **2020**, *45*, 484–502. [CrossRef]
20. Duan, Y.; Zhang, L.; Fan, X.; Hou, Q.; Hou, X. Smart city oriented Ecological Sensitivity Assessment and Service Value Computing based on Intelligent sensing data processing. *Comput. Commun.* **2020**, *160*, 263–273. [CrossRef]
21. Burghardt, K.; Uhl, J.H.; Lerman, K.; Leyk, S. Road network evolution in the urban and rural United States since 1900. *Comput. Environ. Urban Syst.* **2022**, *95*, 101803. [CrossRef]

22. Bergen, K.M.; Loboda, T.; Newell, J.P.; Kharuk, V.; Hitztaler, S.; Sun, G.; Johnson, T.; Hoffman-Hall, A.; Ouyang, W.; Park, K.; et al. Long-term trends in anthropogenic land use in Siberia and the Russian Far East: A case study synthesis from Landsat. *Environ. Res. Lett.* **2020**, *15*, 105007. [CrossRef]
23. Chi, X.; Sartor, M.A.; Lee, S.; Anurag, M.; Patil, S.; Hall, P.; Wexler, M.; Wang, X.-S. Universal concept signature analysis: Genome-wide quantification of new biological and pathological functions of genes and pathways. *Brief. Bioinform.* **2020**, *21*, 1717–1732. [CrossRef] [PubMed]
24. O'Brien, M.J. Genes, memes, and human history: Darwinian archaeology and cultural evolution. *J. Anthropol. Res.* **2003**, *59*, 380–381. [CrossRef]
25. Stent, G.S. You can take the ethics out of altruism but you can't take the altruism out of ethics. *Hastings Cent. Rep.* **1977**, *7*, 33–36. [CrossRef] [PubMed]
26. Pakendorf, B.; Stoneking, M. The genomic prehistory of peoples speaking Khoisan languages. *Hum. Mol. Genet.* **2021**, *30*, R49–R55. [CrossRef] [PubMed]
27. Holderegger, R.; Wagner, H.H. Landscape genetics. *Bioscience* **2008**, *58*, 199–207. [CrossRef]
28. Manel, S.; Schwartz, M.K.; Luikart, G.; Taberlet, P. Landscape genetics: Combining landscape ecology and population genetics. *Trends Ecol. Evol.* **2003**, *18*, 189–197. [CrossRef]
29. Anderson, C.D.; Epperson, B.K.; Fortin, M.-J.; Holderegger, R.; James, P.M.A.; Rosenberg, M.S.; Scribner, K.T.; Spear, S. Considering spatial and temporal scale in landscape-genetic studies of gene flow. *Mol. Ecol.* **2010**, *19*, 3565–3575. [CrossRef]
30. Hu, Z.; Liu, P. The conceptual model and characterizations of landscape genome maps of traditional settlements in China. *Acta Geogr. Sin.* **2015**, *70*, 1592–1605.
31. Hu, Z.; Liu, P.; Shen, X.; Deng, Y.; Li, B. Expression Mechanism of Landscape-Gene Information Unit of Traditional Settlement. *Geogr. Geo-Inf. Sci.* **2010**, *26*, 96–101.
32. Yin, L.; Liu, P. Type expression and spatial differentiation of planar archetype genes of traditional dwellings of the Xiangjiang River Basin. *Geogr. Res.* **2023**, *42*, 2191–2210.
33. Yin, Z.; Li, J. Identification and mapping of landscape gene in historic and cultural villages and towns: A case study of Dayuwan village, Huangpi district. *City Plan. Rev.* **2023**, *47*, 97–104, 114.
34. Chen, Q.; Yang, J.; Luo, S.; Sun, D. Identification and Extraction of the Linpan Culture Landscape Gene. *Trop. Geogr.* **2019**, *39*, 254–266.
35. Li, B.; Li, Z.; Liu, P.; Dou, Y. Activation path of human settlement environment of traditional villages from the perspective ofdouble repair: A case study of Zhang-Guying village in Hunan province. *Geogr. Res.* **2020**, *39*, 1794–1806.
36. Liu, P. On Construction and Utilization of Chinese Traditional Settlements Landscape's Genetic Map. Ph.D. Thesis, Peking University, Beijing, China, 2011.
37. Shen, X.; Liu, P.; Deng, Y.; Wang, L. Landscape communities of traditional settlement in southern China and their values of conservation and utilization. *Geogr. Res.* **2006**, *25*, 485–494.
38. Hu, Z.; Liu, P.; Deng, Y.; Zheng, W. A Novel Method for Identifying and Separating Landscape Genes from Traditional Settlements. *Sci. Geogr. Sin.* **2015**, *35*, 1518–1524.
39. Hu, Z. Retrospects and Outlooks on the Research of Cultural Landscape Gene of Traditional Settlements. *Geogr. Geo-Inf. Sci.* **2024**, 1–9. Available online: http://kns.cnki.net/kcms/detail/13.1330.P.20240612.0904.002.html (accessed on 2 August 2024).
40. Li, X.; Huang, H.; Wang, S. Study on the Identification and Application of Landscape Gene in the Promotion and Construction of Local Cultural Landscape. *Chin. Landsc. Archit.* **2022**, *38*, 29–34.
41. Liu, P.; Liu, Y.; Yang, L.; Li, X. Digital Transmission of Traditional Village Landscape Gene and Its Tourism Value Promotion: A Case Study of Zhangguying Village. *Econ. Geogr.* **2022**, *42*, 232–240.
42. Li, X.; Lei, A. Construction of Traditional Village Cultural Landscape Spatial Database from the Perspective of Landscape Gene: A Case Study of Gaochun District, Nanjing. *J. Hum. Settl. West China* **2023**, *38*, 110–116.
43. Wang, Z.; Zhang, Q.; Chen, Q. Spatial Reconstruction of Tourism-based Traditional Village Settlements and Rural Revitalization: A Case Study of Chongmudang Village in Xuefeng Mountain. *Econ. Geogr.* **2023**, *43*, 229–238.
44. Zhai, Z.; Chang, F.; Li, T.; Li, J.; Shi, Y.; Wu, P. Research on the Cultural Heritage Landscape Genome Maps of Traditional Villages in Shaanxi Province. *Geogr. Geo-Inf. Sci.* **2018**, *34*, 87–94, 113.
45. Zhang, Z.; Dong, G.; Wang, Y. Research on Local Knowledge Graph of Traditional Village Landscape in Dangjia Village, Hancheng City, Shaanxi Province. *Urban Stud.* **2021**, *28*, 25–31.
46. Zhang, P.; Lin, H. *Ancient Fujian Towns: Ancient Rural Commodity Markets in Fujian and Taiwan*; Fujian Map Publishing House: Fuzhou, China, 2008.
47. The Collaborating Organization for This Reference Is the Ministry of Housing and Urban-Rural Development of the People's Republic of China. The Seventh Batch of Chinese Historical and Cultural Towns and Villages List. Available online: https://www.gov.cn/zhengce/zhengceku/2019-09/27/content_5434160.htm (accessed on 18 April 2023).
48. Li, L. Establishment of Characteristic Atlas of Traditional Settlements Landscape in Multicultural Area: Take Southern Shaanxi for Example. *Areal Res. Dev.* **2020**, *39*, 175–180.
49. The Collaborating Organization for This Reference Is the Ministry of Housing and Urban-Rural Development of the People's Republic of China. Guiding Opinions on Effectively Strengthening the Protection of Traditional Chinese Villages. Available online: https://www.mohurd.gov.cn/gongkai/zhengce/zhengcefilelib/201409/20140912_218993.html (accessed on 20 May 2023).

50. Dawkins, R.L. The Selfish Gene. In *Evolution*; Oxford University Press: Oxford, UK, 1976; Volume 32, pp. 245–260.
51. Ministry of Natural Resources of the People's Republic of China. Classification and Codes for Fundamental Geographic Information Feature. Available online: https://std.samr.gov.cn/gb/search/gbDetailed?id=DD3D95E5C18271EBE05397BE0A0AF33F (accessed on 28 May 2023).
52. Wei, N.; Feng, X. Genetic Information Coding and Mapping of Traditional Settlement Landscapes in Northern Shaanxi. *South Archit.* **2023**, 88–97. [CrossRef]
53. Zhou, T.; Long, Y.; Tang, G.; Hu, L. Research on the Hybrid Index Structure for Aggregated Spatial Data. *Geogr. Geo-Inf. Sci.* **2010**, *26*, 7–10.
54. Zhang, G. A Logical Analysis of the Traditional Settlement Living Space from the Perspective of Syntax: Taking Linpu Village as an Example. *J. Minjiang Univ.* **2021**, *42*, 96–104. [CrossRef]
55. Geospatial Data Cloud. Available online: https://www.gscloud.cn/ (accessed on 14 October 2022).
56. Bureau, Fuzhou Natural Resources and Planning Bureau Fuzhou Historical and Cultural City Protection Plan (2021–2035). Available online: https://zygh.fuzhou.gov.cn/zwgk/ghjh/tdlyztgh/202209/t20220923_4439934.htm (accessed on 22 May 2023).
57. Zhang, T. *Historical and Cultural City, Town, and Village Chronicles of Fuzhou City*; Haichao Photography Art Publishing House: Fuzhou, China, 2004.
58. Fujian Provincial People's Government. Preservation Plan for Linpu Village in Chengmen Town, Cangshan District, Fuzhou City, a Historical and Cultural Village of Fujian Province. Available online: https://zygh.fuzhou.gov.cn/zfxxgkzl/gkml/ywgz/cxgh/201903/t20190328_2791942.htm (accessed on 13 May 2023).
59. Hu, H.; Hu, Z.; Wang, F.; Yi, Z. Features of Gene Information Chain of Traditional Village Landscape and Its Identification. *Econ. Geogr.* **2019**, *39*, 8.
60. Zhu, H.; Liu, J.; Wang, L.; Sun, D. Influence factors of residents' attitude and support to tourism development in ancient and cultural village: A case study of Linpu. *World Reg. Stud.* **2018**, *27*, 166–176.
61. Yang, L.; Hu, Y. Production and Inheritance of Intangible Cultural Landscape Gene in Traditional Villages: A Case Study of Huangdu Village in Tongdong Autonomous County. *Econ. Geogr.* **2022**, *42*, 208–215.
62. Kuroda, N. Conservation Design for Traditional Agricultural Villages: A Case Study of Shirakawa-go and Gokayama in Japan. *Built Herit.* **2019**, *3*, 7–23. [CrossRef]
63. Cai, G.; Xu, L.; Gao, W.; Hong, Y.; Ying, X.; Wang, Y.; Qian, F. The Positive Impacts of Exhibition-Driven Tourism on Sustainable Tourism, Economics, and Population: The Case of the Echigo-Tsumari Art Triennale in Japan. *Int. J. Environ. Res. Public Health* **2020**, *17*, 1489. [CrossRef] [PubMed]

buildings

MDPI

Article

Development of a Building Evaluation Framework for Biophilic Design in Architecture

Rafa Ramisha Tabassum and Jihyun Park *

Department of Architecture, Ewha Womans University, 52 Ewhayeodae-gil, Seodaemun-gu, Seoul 03760, Republic of Korea; ramisha.rafa@ewhain.net
* Correspondence: jh.park@ewha.ac.kr

Abstract: Biophilic design has gained prominence in architecture as a strategy to integrate natural elements into built environments, aiming to improve human comfort, well-being, and environmental sustainability. This paper critically reviews existing biophilic design frameworks, identifying gaps in their integration of health outcomes and the relationship between indoor and outdoor environments, with a primary focus on indoor environmental quality (IEQ) factors such as air quality, natural light, thermal comfort, and acoustics. The review was conducted systematically, comparing frameworks using criteria such as health benefits, human–nature connections, and user satisfaction. Key findings highlight a lack of comprehensive frameworks that explicitly link biophilic design to measurable improvements in human health, comfort, and satisfaction. To address this gap, the study proposes an enhanced biophilic design framework that bridges IEQ-focused indoor environments with urban outdoor elements, prioritizing human comfort alongside environmental sustainability. The framework also emphasizes the role of post-occupancy evaluation (POE) in assessing the real-world performance of biophilic interventions, particularly in achieving user satisfaction and ensuring long-term effectiveness. This research contributes to the advancement of biophilic design by offering actionable strategies for architects, urban planners, and policymakers to adopt biophilic principles that create resilient, healthy, and sustainable spaces. Additionally, the study emphasizes the need for empirical testing and validation of the proposed framework, including POE and user surveys, to assess its real-world impact on human well-being and satisfaction, thus setting the stage for future research.

Keywords: biophilic design; indoor environmental quality; human comfort; urban sustainability; environmental features; nature integration; well-being in architecture; sustainable urban development; human–nature connection; health and well-being; resilient environments

1. Importance of Biophilic Design in Architecture

Biophilic design has emerged as a crucial approach in sustainable architecture, emphasizing the integration of natural systems into the built environment to enhance both human well-being and ecological sustainability [1–5]. By embedding natural elements into urban structures, biophilic design promotes spaces that are not only ecologically responsible but also psychologically fulfilling [6–8]. This approach effectively addresses urban challenges such as the heat island effect by increasing vegetation cover, shading, and evapotranspiration. These strategies reduce environmental impact while enhancing human comfort and satisfaction [9,10].

This disconnect between nature and urban settings is evident in various fields, including education, healthcare, architecture, and urban development. This underscores the need to design buildings that not only respond to sustainability challenges but also positively contribute to the surrounding environment [6,11–15].

As global climate change continues to intensify, there is a growing need to reevaluate traditional architectural practices. Newer methods must contribute positively to sustainability while prioritizing human comfort [16–18]. In this context, biophilic design stands out compared to other sustainable architectural concepts such as Green Architecture, Climate-Responsive Architecture, Net-Zero Architecture, and Vernacular Architecture (see Table 1). These approaches each address specific aspects of sustainability; however, biophilic architecture provides a more holistic solution for fostering long-term resilience in urban environments [19–25].

Table 1. A Comparative review of architectural concepts that aim to achieve sustainability.

		Architecture with the Concepts of Sustainability				
		Green Architecture	**Climate-Responsive Architecture**	**Net-Zero Architecture**	**Vernacular Architecture**	**Biophilic Architecture**
Sustainable Components	Durability	●	◍	●	◍	●
	Indoor Air Quality (IAQ)	●	●	●	●	●
	Health and Well-being	◍	◍	◍	●	●
	Energy and CO_2 Emissions	●	●	●	◍	●
	Sustainable Materials	●	●	●	●	●
	Water Conservation	●	○	○	●	●
	Surface-water Run-off Pollution	○	○	○	○	●
	Waste Reduction	●	○	●	○	●
	Management	●	◍	●	◍	●
	Ecology	○	○	○	○	●

● Fully integrated. ◍ Partially considered with limitations. ○ Not emphasized.

A key aspect of this reassessment involves focusing on indoor environmental quality (IEQ), which plays a vital role in optimizing occupant health, comfort, and productivity [26–29]. Given that modern humans spend approximately 90% of their time indoors [30], the quality of indoor environments—particularly in the integration of natural elements—becomes pivotal to overall well-being [8,14]. While biophilic design shows promise in addressing these concerns, more empirical studies are needed to validate its impact on health outcomes [5,8,14,31]. By addressing the needs people seek to fulfill through their built environments, architects can create spaces that are both sustainable and ecologically regenerative [3,27,32,33].

While other approaches, such as Green Architecture, emphasize energy efficiency and material sustainability, or focus on adapting buildings to climate conditions, like Climate-Responsive Architecture, biophilic architecture is distinguished by its holistic integration of sustainable components. These include durability, indoor air quality (IAQ), health and well-being, energy and CO_2 emissions, sustainable materials, water conservation, surface-water run-off pollution, waste reduction management, and ecology [3,34–38]. This broad focus on both environmental and human well-being sets biophilic architecture apart from other sustainable design concepts. For instance, Green Architecture often prioritizes measurable environmental outcomes like CO_2 emission reduction, while biophilic architecture focuses more holistically on integrating natural elements to enhance the emotional and psychological comfort of occupants [11,39–41]. Moreover, when compared to Net-Zero Architecture, which is primarily driven by energy efficiency and minimizing CO_2 emissions, biophilic architecture takes a broader approach by addressing not only energy and environmental performance but also the human experience within the space [3,42–44]. Additionally, while Vernacular Architecture is grounded in local materials and traditional techniques, biophilic architecture incorporates ecological principles to foster a strong connection with nature, addressing both the physical and emotional dimensions of the human experience in buildings [31,43–45].

The need for beneficial contact with nature is well documented, with substantial evidence linking it to improved health and well-being. However, achieving this in the built environment presents significant challenges, particularly in ensuring adequate exposure to natural light, ventilation, natural materials, and green spaces [6,8]. This disconnection has profound implications, not only for individual well-being but also for the broader goal of environmental sustainability. Therefore, biophilic design addresses these gaps by offering a more integrated approach that harmonizes indoor and outdoor experiences, particularly through elements such as natural light, ventilation, and green spaces. This integration improves the overall quality of life for building occupants [10,11,46–49].

The biophilic framework proposed in this study aims to fill the gaps left by existing frameworks, particularly by integrating both indoor and outdoor environmental factors. Existing approaches may excel in specific areas, such as energy conservation in Net-Zero Architecture or cultural responsiveness in Vernacular Architecture, but biophilic architecture incorporates a broader range of sustainable components, making it a comprehensive and flexible solution for urban development [50,51].

In conclusion, biophilic design not only addresses the pressing issues of sustainability but also enhances the quality of life for building occupants by fostering a stronger connection to nature. Its holistic approach ensures that the built environment caters to both ecological and human needs, making it an essential framework for the future of sustainable architecture [31].

2. Materials and Methods

This study employed a structured methodology to develop a comprehensive biophilic design framework, focusing on human comfort, well-being, and environmental sustainability, with a slight inclusion of outdoor elements. The methodology was divided into two key phases: (1) a systematic literature review and (2) framework development.

2.1. Systematic Literature Review

The first phase of the methodology involved a systematic review of existing biophilic design frameworks, particularly in architectural and urban planning contexts. The aim was to identify key gaps and limitations, particularly regarding human health, well-being, and environmental sustainability. This process followed a rigorous structure, detailed below.

Search Strategy

The literature search targeted peer-reviewed articles, books, and authoritative reports on biophilic design using databases such as Scopus, Web of Science, and Google Scholar. Keywords included "biophilic design", "human comfort", "well-being in architecture", "sustainable urban development", "nature integration", and "Indoor Environmental Quality (IEQ)". A particular focus was placed on frameworks that directly connected biophilic elements with measurable impacts on human health and environmental performance.

Inclusion/Exclusion Criteria

The following criteria guided the inclusion of studies:

- Published between 1980 and 2024.
- Direct focus on biophilic design frameworks in urban and architectural settings.
- Empirical or theoretical analysis of the impact of biophilic design on human health, well-being, or environmental outcomes.
- Studies focusing solely on natural landscapes or unrelated to built environments were excluded, ensuring the review remained relevant to architectural applications and human comfort.

Data Extraction and Synthesis

Key data were extracted on biophilic elements (e.g., natural light, vegetation, air quality), methodologies for assessing their impact, and their outcomes related to human comfort, health, and environmental performance. The comparative analysis of existing frameworks highlighted strengths, weaknesses, and areas for improvement, particularly in connecting indoor and outdoor environments and linking biophilic elements to human health benefits. Specific attention was given to frameworks that had measurable outcomes, addressing the gap identified in earlier studies.

2.2. Development of the Biophilic Design Framework

Building on the findings from the systematic literature review, this study developed an enhanced biophilic design framework, addressing gaps in the current frameworks. This phase focused on three primary areas.

Human Comfort and Well-being

The new framework prioritized measurable outcomes related to human health and comfort, including metrics such as visual comfort, air quality, stress reduction, and thermal comfort. Key biophilic design elements—such as natural ventilation, daylighting, green spaces, and water features—were integrated to enhance these outcomes in both indoor and outdoor environments.

Indoor–Outdoor Connection

A key feature of the proposed framework was the seamless integration of indoor and outdoor environments, promoting a holistic connection with nature. The framework incorporated design strategies like open spaces, balconies, rooftop gardens, and visual access to natural surroundings, aiming to foster a stronger connection between occupants and nature, thereby enhancing well-being.

Sustainability and Practical Considerations

The framework also addressed the practical challenges of biophilic design, including cost-effectiveness, scalability, and maintenance. It proposed flexible solutions that could be adapted to a range of urban settings, from residential to commercial spaces. Sustainable materials, water conservation, energy efficiency, and waste management strategies were emphasized to support the environmental sustainability component of biophilic design.

Framework Evaluation: To ensure the robustness of the proposed biophilic design framework, this study recommends empirical validation through future research, using methods such as case studies, user surveys, or experimental evaluations. These tests would assess the practical applicability of the framework across diverse built environments and its effectiveness in improving human comfort, satisfaction, and environmental performance.

3. Review of Existing Biophilic Design Frameworks

3.1. Historic Context and Evolution

The concept of biophilic design, which has been gaining increasing attention in recent architectural discourse, has deep historical roots, tracing back to humanity's inherent connection with nature. Before the industrialization and urbanization eras, human societies relied heavily on interaction with natural environments for both survival and well-being. Historically, human settlements were multi-sensory environments, where natural elements such as light, air, greenery, water, and sound played a central role in daily life [10,52,53]. These natural settings provided essential experiences that shaped emotional and psychological well-being, a connection that neuroscientific studies now confirm as a fundamental human need [54–56]. Research employing modern neuroimaging techniques has shown that contact with nature positively influences brain activity, reducing stress and promoting cognitive restoration [54–56].

The term "Biophilia", foundational to biophilic design, originates from the Greek words "philia" (love) and "bios" (life or living things). Introduced by German psychoanalyst Erich

Fromm in 1973, it describes humanity's inherent "love for life" and attraction to the natural world. This concept was popularized by American biologist Edward O. Wilson in his seminal work *Biophilia* (1984), which introduced the "Biophilia Hypothesis" [6,50]. This theory posits that humans possess an innate affinity for living systems, shaped by evolutionary processes that foster psychological and physical well-being through close interactions with nature [57–61]. Wilson's hypothesis has catalyzed further philosophical and scientific inquiries into how reconnecting with nature can enhance human health, comfort, and satisfaction.

Despite its formal recognition in the late 20th century, biophilic design is not a novel architectural concept. The integration of natural forms and elements into built environments has been prevalent throughout history, long before the term "biophilia" existed. The Art Nouveau movement of the late 19th century, for example, drew heavily on organic and natural motifs, with architects and designers like Victor Horta, Hector Guimard, Louis Comfort Tiffany, and Antonio Gaudí creating works inspired by the natural world. Their use of flowing lines, plant-like structures, and biomorphic forms served not only as decorative features but also aimed to harmonize human-built environments with nature, fostering a deeper emotional connection to the surrounding world [5,17].

Biophilic design, as it is understood today, represents a re-emergence of this long-standing human tradition of integrating natural elements into architecture. From ancient civilizations to the Art Nouveau movement and contemporary biophilic design, the use of natural forms in architecture has persisted as a means of promoting human well-being. This historical evolution underscores the critical importance of biophilic principles in modern architecture, particularly in urban contexts where nature is often absent. By integrating natural elements into urban spaces, biophilic design offers a path to enhancing human comfort, satisfaction, and health in environments increasingly detached from nature due to urbanization and technological advancements [26,62,63].

In conclusion, the historical context of biophilic design reveals a continuous effort to reconnect built environments with the natural world. As urbanization continues to reshape human living spaces, the re-integration of biophilic elements emerges not only as an aesthetic choice but as a necessary design strategy for promoting human well-being and environmental sustainability. This evolution demonstrates that biophilic design is not merely a modern trend but a fundamental design principle central to architectural practices throughout history [5,10,64].

3.2. Analysis of Biophilic Design Frameworks

Biophilic design frameworks have evolved over time, offering diverse methods to integrate natural elements into built environments with the overarching goal of enhancing human comfort, health, and well-being. This section presents a comparative analysis of seven prominent frameworks, detailing their focus areas, key dimensions, and strategies [11,65,66].

Table 2 outlines the main features of each framework, highlighting how they approach biophilic design and human–nature interaction in architectural and urban settings. The frameworks range from Kellert et al.'s focus on environmental features and natural forms to Ryan et al.'s emphasis on multi-sensory biophilic interventions. While each framework brings unique strengths, there are variations in how they address key design patterns and human well-being [11,12,32,66].

A comprehensive review of existing biophilic design frameworks reveals significant gaps, especially regarding their application to urban environments and the explicit connection between design elements and health benefits. While these frameworks emphasize biophilic principles, they often lack clarity in defining how specific urban design elements correlate with health improvements, such as enhanced air quality, stress reduction, and cognitive function [3,67].

Table 2. Comparative analysis of biophilic design frameworks.

Research	Main Focus	Dimensions/Attributes	Key Patterns/Strategies
Kellert et al. [2]	Connection between humans and nature via architecture	6 attributes: Environmental features, Natural shapes and forms, Natural patterns, Light, Place, Evolutionary perspectives	3 types of experience: direct, indirect, symbolic interaction with nature
Ryan and Browning (14 Patterns) [39]	Framework to enhance human well-being in the built environment	14 patterns: Visual connection, Non-visual stimuli, Airflow variability, etc.	Strategies for sensory experiences, restorative environments
Kellert & Calabrese (Experiences) [6]	Expands on Kellert's 2008 work by detailing direct experiences with nature	Direct (water, plants, light), Indirect (images, colors, shapes), Experience of place	Comprehensive link between environment and human interaction
Makes and Sense (Economic Benefits of Biophilic) [65]	Stress reduction, cognitive function improvement	4 major areas: Nature in space, Natural analogs, Nature of space	Patterns for sensory balance, space configuration
Brielmann et al.(Fractal Effects in Urban Design) [55]	Neuroscience of human comfort and aesthetics	Biophilic elements based on cognitive response to the environment	Principles related to fractals, complexity, and mental stimulation
Ryan and Browning, (Seven Dimensions) [18]	Framework focused on enhancing health and well-being through biophilia	7 dimensions: Visual, Non-visual, Auditory, Olfactory, Tactile, Thermal, Spatial	Multi-sensory biophilic interventions
Untaru et al. (Biophilic Interventions) [66]	Biophilic design for urban environments	Urban biophilic interventions for well-being and sustainability	Focus on green infrastructure, urban integration

For example, design elements like green courtyards, rooftop gardens, and water features are frequently recommended but without a clear articulation of their direct health benefits. This omission makes it difficult for practitioners to understand how these elements can be strategically implemented to maximize both physical and mental health outcomes [68–72]. As a result, practitioners may struggle to implement biophilic design strategies that deliver tangible health benefits, particularly in complex urban environments where these considerations are most crucial [42,73].

By explicitly linking biophilic design elements to their health benefits, as demonstrated in Table 3, this structured approach enhances the practical application of biophilic principles in urban environments. For instance, incorporating plants and natural ventilation into buildings not only improves air quality but also regulates humidity and reduces noise, contributing to overall well-being [63,74]. Similarly, green streets and pedestrian-friendly spaces foster a sense of connection with nature, improve air quality, and alleviate stress—key for urban dwellers [5,75–77].

Moreover, this structured approach addresses the need for a more user-friendly framework, where each category is clearly organized and presented. Such organization provides a comprehensive view of how biophilic design can be strategically implemented across various urban scales, ultimately benefiting the well-being of city inhabitants [8,78,79].

To fully unlock the potential of biophilic design in urban settings, future frameworks must provide clear, actionable strategies that link design elements to health outcomes. By doing so, architects and urban planners can create sustainable, resilient environments that support human well-being and ecological balance [10,80].

Table 3. Integration of biophilic design elements and health benefits into urban design components.

Urban Design Components	Biophilic Design Elements	Health Benefits
Buildings	- Green Courtyards and Roofs	- Better Air Quality - Humidity Regulation - Noise Reduction- Stress Reduction - Enhanced Cognitive Function
	- Indoor and Outdoor Plants	- Better Air Quality - Stress Reduction - Enhanced Immune System
	- Water Features/Reservoirs	- Humidity Regulation - Stress Reduction - Faster Recovery
	- Natural Daylight and Ventilation	- Better Air Quality - Stress Reduction
	- Nature Images and Paintings	- Stress Reduction - Improved Life Satisfaction
Public Spaces	- Green Areas and Pocket Parks	- Stress Reduction - Enhanced Cognitive Function—Sense of Connection
	- Ecology Parks and Native Species Yards	- Enhanced Immune System - Faster Recovery
	- Water Features/Reservoirs	- Stress Reduction - Humidity Regulation
	- Natural Sounds/Creating Artificial Nature Sounds	- Stress Reduction
Streets	- Green Streets and Road-side Parks	- Better Air Quality - Sense of Connection
	- Skinny Streets and Vegetated Swales	- Better Air Quality - Noise Reduction
Transport	- Accessibility/Pedestrian Connections	- Stress Reduction - Sense of Connection
	- Proper Road, Rail, Bicycle, and Pedestrian Networks	- Improved Life Satisfaction
Landscape	- Native and Edible Landscaping	- Enhanced Immune System - Sense of Connection
	- Thermally Treated Timber and Natural Materials	- Enhanced Cognitive Function—Improved Life Satisfaction

3.2.1. Strengths and Limitations Regarding IEQ and POE

In this section, we evaluate the strengths and limitations of these frameworks (Table 2) concerning indoor environmental quality (IEQ) and post-occupancy evaluation (POE), two critical areas for assessing building performance and human comfort.

a. Kellert et al. (Attributes) [2]

Strengths: The framework emphasizes enhancing human well-being through natural elements and evolutionary perspectives. This indirectly supports IEQ by improving the emotional and cognitive experience of users through natural light, air quality, and access to views.

Limitations: While it touches on environmental features, it does not specifically address IEQ parameters like thermal comfort, acoustics, or air quality in depth, nor does it incorporate POE for evaluating user satisfaction post-occupancy.

b. Ryan and Browning (14 Patterns) [39]

Strengths: This framework, with its 14 patterns, is more explicit about improving aspects related to IEQ, such as thermal variability, airflow, and natural ventilation, aligning directly with physical comfort factors. Some patterns also implicitly support POE through adaptability in design.

Limitations: Although it improves certain IEQ aspects, the framework doesn't offer specific guidance on post-occupancy evaluation to assess the success of these interventions over time.

c. Kellert and Calabrese (Experiences) [6]

Strengths: Their framework integrates direct and indirect experiences of nature, fostering environments that can enhance sensory experiences tied to air quality, lighting, and thermal comfort.

Limitations: The framework lacks explicit mention of IEQ standards or a structured approach to conducting POE to measure how well the space performs for users after occupancy.

d. Makes and Sense (Economic Benefits of Biophilic) [65]

Strengths: This work focuses on stress reduction and cognitive function, with some consideration of acoustic comfort and natural ventilation, tying into IEQ elements. This framework could be adapted for ongoing user evaluation via POE.

Limitations: Although it touches on comfort-related aspects of IEQ, the framework does not comprehensively cover all IEQ factors (e.g., thermal comfort and lighting quality) or provide detailed mechanisms for POE.

e. Brielmann et al. (Fractal Effects in Urban Design) [55]

Strengths: Salingaros connects biophilia to neuroscience, emphasizing mental well-being and cognitive engagement with space, which supports the psychological dimension of IEQ.

Limitations: Physical IEQ aspects like air quality and thermal conditions are not explicitly addressed. There is also no mention of post-occupancy evaluation to validate user satisfaction or comfort.

f. Ryan and Browning(Seven Dimensions) [18]

Strengths: This framework is particularly strong in addressing multi-sensory biophilic interventions, including non-visual sensory stimuli that align with IEQ. It focuses on auditory, thermal, and spatial experiences, making it relevant for enhancing indoor comfort.

Limitations: While it addresses IEQ aspects, it lacks formalized methods for evaluating the success of these interventions through POE, leaving a gap in post-occupancy assessment.

g. Untaru et al. (Biophilic Interventions) [66]

Strengths: Han's framework is more urban-focused but includes interventions such as green infrastructure and urban greenery, which can contribute to improving outdoor IEQ. It can also incorporate POE through community engagement and satisfaction metrics in urban settings.

Limitations: The framework lacks detailed strategies for evaluating indoor IEQ in buildings and does not provide structured POE techniques to track ongoing user satisfaction.

3.2.2. Identified Key Gaps

a. IEQ Gaps: Many frameworks acknowledge sensory aspects, such as natural light, air quality, and temperature regulation, but fail to address IEQ comprehensively, particularly with regard to acoustics, thermal comfort, and detailed indoor air quality standards. More integration of IEQ standards such as the ASHRAE or the WELL Building Standard® could strengthen their applicability [3,62].
b. POE Gaps: None of the frameworks explicitly includes methods or tools for post-occupancy evaluation, which is a critical gap. POE could provide valuable feedback on how biophilic interventions perform in real-world settings, guiding future improvements and ensuring sustained user comfort and satisfaction [8,31,81].
c. Recommendations for Improvement

Integrate Comprehensive IEQ Standards

Frameworks should incorporate more robust IEQ standards, ensuring a holistic approach to indoor environments that considers acoustics, thermal comfort, and air quality. The incorporation of standards from organizations such as the ASHRAE or the WELL Building Standard® would enhance the frameworks' effectiveness and applicability in various settings [3].

Establish Structured POE Mechanisms

There is a critical need for frameworks to include methods for conducting structured POE, which can help in assessing user comfort and satisfaction after occupancy. By integrating feedback mechanisms and evaluation tools, these frameworks could be continuously improved based on real-world performance and user feedback [8].

Strengths

The frameworks excel in integrating nature-based design strategies that promote well-being, stress reduction, and sensory engagement. These contribute to the overall comfort of occupants, even if IEQ and POE are not explicitly mentioned.

The use of multi-sensory design patterns is a major strength, as it encourages designers to consider more than just visual aesthetics when integrating biophilia, extending to auditory and tactile experiences.

In conclusion, while these biophilic design frameworks offer valuable approaches to enhancing human–nature interaction in architecture, they often fall short in providing comprehensive guidelines on IEQ and lack structured POE mechanisms to assess user comfort post-occupancy. Expanding these frameworks to include rigorous IEQ standards and incorporating POE would improve their application in contemporary architectural practice.

4. Biophilic Design Aspect for Human Comfort

4.1. Indoor and Outdoor Design Elements

Biophilic design presents a comprehensive approach that integrates natural elements into both indoor and outdoor environments, thereby enhancing occupant well-being and promoting sustainability. Table 4 below summarizes key aspects of indoor and outdoor biophilic design, categorizing their primary foci, key elements, benefits, challenges, and essential considerations for implementation.

Indoor biophilic design emphasizes the integration of natural elements such as plants, natural light, and water features within confined spaces, contributing to improved air quality, stress reduction, and increased productivity [3,37,81–83]. This approach is highly adaptable to various climates and locations, allowing customization based on user preferences and needs [8,53,84,85]. However, challenges such as limited space, reliance on artificial lighting, and the maintenance of indoor plants present notable obstacles [17,53,62,86–88].

Conversely, outdoor biophilic design integrates buildings and open spaces with natural surroundings. This approach factors in landscaping, weather, and biodiversity preservation, which are critical for urban planning and the conservation of green spaces [3,15,75,89–91]. While outdoor designs can be unpredictable due to natural variations, they offer significant benefits, including improved mental well-being, physical health, and community bonding [64,92–94]. These features are particularly valuable in urban settings, where access to nature may be limited [95–98].

Table 4. Aspects of indoor and outdoor biophilic design.

Aspect	Indoor Biophilic Design	Outdoor Biophilic Design
Primary Focus	Enhancing interior spaces	Integrating natural elements into outdoor environments
Key Elements	Living walls, indoor plants, natural light, water features	Green roofs, gardens, parks, natural water features, landscaping
Human Benefits	Improved air quality, stress reduction, increased productivity	Enhanced mental well-being, physical health, and community bonding
Design Integration	Incorporating nature into architecture and interior elements	Landscaping, urban planning, preservation of green spaces
Challenges	Limited space, artificial lighting, maintenance of indoor plants	Weather considerations, biodiversity preservation
Sensory Experience	Aroma, touch, visual appeal, sounds of nature	Sounds of nature, visual aesthetics, tactile experiences
Application Scales	Homes, offices, commercial spaces	Parks, public spaces, urban planning projects
Accessibility	Available in various climates and locations	Dependent on outdoor space availability, weather conditions
Cost Considerations	Initial setup costs, maintenance expenses	Initial landscaping costs, ongoing maintenance, and upkeep
Sustainability Focus	Sustainable materials, energy-efficient designs	Preservation of natural ecosystems, eco-friendly landscaping
Community Impact	Indoor green spaces within communities	Public parks, communal outdoor spaces
Wellness Connection	Mental and emotional well-being, connection to nature indoors	Physical health benefits, community well-being outdoors

4.2. Impact on Psychological and Physiological Comfort

Biophilic design, both indoors and outdoors, has a profound impact on human comfort by addressing psychological and physiological well-being. By connecting occupants with nature, this design approach transcends aesthetic appeal and fulfills the innate human need for natural interaction.

(1) Psychological Comfort: Indoor biophilic design creates calming environments that reduce stress and anxiety. The inclusion of plants, natural light, and water features indoors supports mental well-being, enhances cognitive function, and increases creativity and productivity [3,37,42,44,99,100]. Given that people spend a significant portion of their lives indoors, connecting with nature inside these spaces helps mitigate the disconnect from the natural world experienced by many urban dwellers [101–103].

Outdoor biophilic design, on the other hand, enhances psychological comfort by providing spaces for physical activity, social interaction, and relaxation in natural settings. Exposure to natural sounds, visual aesthetics, and tactile experiences outdoors significantly reduces stress, promotes mental clarity, and improves overall mood [17,26,81,104,105]. This design fosters community bonding by creating accessible natural spaces where people can gather and interact, contributing to collective well-being [5,14,18,40,106].

(2) Physiological Comfort: Biophilic design also positively influences physical health. Indoors, it improves air quality, regulates humidity, and creates more comfortable thermal environments. For example, indoor plants act as natural air purifiers, absorbing toxins and releasing oxygen, which contributes to better respiratory health and overall physical well-being [3,74,107–111]. Natural light and water features support circadian rhythms and reduce dependence on artificial lighting, promoting a healthier living or working environment [5,44,85,112].

Outdoors, biophilic design mitigates urban heat island effects through vegetation and green spaces [5,89,113,114]. Parks and gardens designed with biophilic principles provide essential spaces for exercise, relaxation, and social interaction, all of which contribute to physical well-being. Additionally, the inclusion of natural elements such as water bodies and vegetation in urban settings helps moderate temperatures, reduce pollution, and create a healthier environment for communities [63,64,81,115].

In conclusion, biophilic design integrates nature into both indoor and outdoor spaces, significantly enhancing human comfort by addressing psychological and physiological needs. By connecting people with nature, this design approach not only improves individual well-being but also promotes a healthier, more sustainable built environment.

5. Development of a Biophilic Evaluation Framework

The newly developed framework (Table 5), Integrated Biophilic Design for Human Comfort and Sustainability, addresses critical gaps in previous biophilic design frameworks by combining indoor and outdoor design strategies with a strong focus on human comfort, sustainability, and the integration of indoor environmental quality (IEQ) and post-occupancy evaluation (POE). This framework is unique because it extends beyond traditional indoor-centric approaches by incorporating outdoor dimensions while maintaining a primary focus on indoor spaces, where people spend 90% of their time. It also prioritizes empirical data collection through POE, making it not just a design guideline but a dynamic evaluation tool that can adapt to evolving building environments and occupant needs.

Table 5. Integrated biophilic design for human comfort and sustainability.

Design Dimension	Components	Human Comfort Factors	Sustainability Factors	Integration of IEQ and POE	Performance Metrics
Direct Experiences of Nature	Natural light, plants, water features, outdoor views, fresh air	Visual comfort, air quality, emotional well-being	Energy efficiency, biodiversity, water conservation	IEQ: Focus on natural ventilation, daylight, and air quality. POE: User satisfaction with light levels, air quality, and environmental controls, extended to outdoor views and fresh air.	- Daylight availability (lux) - CO_2 concentration (ppm) - Thermal comfort (°C) - Glare index (DGI) - User satisfaction with natural features (indoor and outdoor)
Indirect Experiences of Nature	Natural materials, views of nature, nature-inspired shapes, green roofs	Cognitive engagement, psychological relaxation	Use of sustainable materials, reduced waste, integration of urban green infrastructure	IEQ: Use of low-VOC, sustainable materials indoors, green roofs for better thermal insulation. POE: Feedback on material quality, engagement with nature (views, greenery).	- VOC concentration (mg/m^3) - Visual satisfaction (survey) - Thermal resistance (R-value) - Survey: engagement with nature (views, indoor plants)
Dynamic and Diffuse Light	Daylight, circadian lighting, shading systems, skylights	Visual comfort, stress reduction, circadian support	Energy reduction, adaptability of lighting systems	IEQ: Focus on circadian lighting, dynamic daylight integration. POE: Impact of lighting on sleep cycles, mood, and productivity (indoors), with some outdoor considerations for shade.	- Circadian stimulus (CS) - Illuminance (lux) - Luminance contrast ratio - Melatonin suppression (%) - Sleep quality surveys (indoor environment focus)
Thermal and Airflow Variability	Adaptive ventilation, outdoor breezeways, natural cooling systems	Thermal comfort, cognitive performance, air quality	Energy efficiency through natural cooling, reduced HVAC demand	IEQ: Enhance thermal comfort with adaptive systems indoors, extend airflow strategies to outdoor spaces. POE: Measure satisfaction with temperature control, airflow indoors and outdoors.	- Air speed (m/s) - Operable windows usage (%) - Survey: thermal satisfaction (indoor and outdoor spaces) - Radiant temperature asymmetry (°C)
Non-visual Sensory Stimuli	Soundscapes, textured materials, natural sounds (water, wind)	Auditory comfort, tactile engagement	Reduced noise pollution, enhanced material durability	IEQ: Focus on acoustic comfort, especially with natural sound elements. POE: Evaluate responses to soundscapes and tactile materials indoors and in semi-outdoor spaces (courtyards, atriums).	- Background noise level (dBA) - Tactile response to materials (survey) - Surface texture coefficient - Sound masking and absorption (indoor and outdoor sound quality)
Spatial Variability and Prospect-Refuge	Enclosed spaces, open vistas, urban gardens, rooftop retreats	Psychological comfort, sense of safety, cognitive restoration	Flexible use of space, efficient land use	IEQ: Offer variety in spatial experiences, including enclosed and open areas, indoors and out. POE: Surveys on comfort and sense of security in enclosed vs. open spaces (indoors/outdoors).	- Occupant density (per m^2) - Privacy score (survey) - Visual connection to outdoors (window-to-wall ratio) - Sense of security and comfort (survey: indoor and outdoor)
Biophilic Urban Integration	Urban green spaces, street trees, water management systems, biodiversity	Outdoor comfort, social interaction, overall well-being	Urban biodiversity, stormwater management, pollution control	IEQ: Leverage green infrastructure for better air quality and outdoor comfort. POE: Community engagement and feedback on outdoor spaces, integration with the indoor environment.	- Pollution reduction (PM2.5/PM10) - Stormwater retention rate (mm/hr) - Urban biodiversity index (species count) - Satisfaction with outdoor spaces (surveys)

5.1. Development of Biophilic Framework and Key Strengths

(1) Holistic Integration of Indoor and Outdoor Environments

One of the most significant strengths of this framework is its inclusion of both indoor and outdoor biophilic design elements. Many existing frameworks focus primarily on indoor spaces, addressing direct and indirect experiences with nature, but often fall short in connecting these experiences with the outdoor environment. The current table bridges this gap by including outdoor strategies like urban gardens, green roofs, and outdoor views, which improve both psychological well-being and environmental performance [3].

This holistic approach is particularly innovative because it recognizes that outdoor spaces—such as rooftop gardens, green walls, or tree-lined walkways—can be as vital as indoor spaces in promoting well-being [31]. The focus on urban integration through features like biodiversity and stormwater management positions this framework as more environmentally comprehensive than many earlier models [31].

(2) Detailed Focus on Human Comfort and Well-being

Unlike earlier frameworks that are more general in their approach to human comfort [52], this table is explicitly geared towards ensuring various forms of comfort—visual, thermal, acoustic, and psychological—while also addressing cognitive and emotional needs. For instance, the Direct Experiences of Nature dimension focuses on comfort factors such as visual comfort and emotional well-being, supported by specific design interventions like natural lighting and air quality improvement [5].

What sets this table apart is the incorporation of non-visual sensory stimuli and prospect-refuge theory to create a more inclusive and adaptable environment. Non-visual sensory experiences, such as soundscapes or tactile textures, go beyond traditional frameworks that focus predominantly on the visual connection to nature [64]. These deeper sensory experiences are crucial in reducing stress and increasing occupant satisfaction, as highlighted in the table's Non-Visual Sensory Stimuli dimension.

(3) Rigorous Performance Metrics

A key differentiator of this framework is its extensive performance metrics, which are designed to be measurable and evidence-based. Existing frameworks provide qualitative strategies for implementing biophilic design, but they often lack detailed metrics for measuring success. By contrast, this table incorporates quantifiable metrics such as daylight availability (lux), CO_2 concentration (ppm), and circadian stimulus (CS). These metrics allow for the empirical evaluation of biophilic design outcomes, making it more robust and adaptable to future innovations in design and technology.

The emphasis on post-occupancy evaluation (POE) is another unique strength. POE allows for continuous feedback from occupants, enabling designers and architects to adapt the built environment to real-time needs and preferences. While POE is mentioned in other frameworks, it is often underutilized or lacks specificity [81,116]. This table, however, makes POE central by including user satisfaction surveys related to various comfort factors and performance metrics.

(4) Sustainability and IEQ Synergy

This table also successfully integrates sustainability factors with Indoor Environmental Quality (IEQ), something many older frameworks treat as separate or secondary. The Dynamic and Diffuse Light dimension, for example, focuses on the energy efficiency of lighting systems while simultaneously promoting circadian rhythm support for human health. This dual focus ensures that design choices do not just meet environmental targets but also contribute directly to occupant comfort and well-being [3,64].

Additionally, the integration of green infrastructure (such as green roofs or urban gardens) in the Biophilic Urban Integration dimension emphasizes sustainability in a way that surpasses previous models. Urban integration is also present in other frameworks, but this table adds a stronger connection between indoor and outdoor spaces by tying stormwater management, air pollution reduction, and urban biodiversity directly to human comfort outcomes, ensuring that sustainability goals do not compromise occupant well-being [31].

5.2. Comprehensive Biophilic Design Factors for Human Comfort and Satisfaction

This section introduces a newly developed biophilic design framework aimed at enhancing human comfort and satisfaction. Building on existing frameworks, this new model integrates insights from previous analysis sections while addressing identified gaps in current biophilic design approaches. The framework's goal is to offer a more comprehensive solution that effectively targets multiple dimensions of human comfort and satisfaction within both architectural and urban spaces.

The framework was constructed by synthesizing key findings from Table 2 (Comparative analysis of biophilic design frameworks) and Table 6 (Integration of biophilic design elements and health benefits into urban design components). It categorizes biophilic elements into factors that contribute to human comfort and satisfaction while presenting both direct and indirect biophilic experiences in architecture. Additionally, the framework suggests specific biophilic elements that can be implemented to improve well-being [17,18,117].

Table 6. Comprehensive biophilic design Factors for human comfort and satisfaction.

Human Comfort and Satisfaction Factors	Direct and Indirect IntegrationExperiences in Architecture	Biophilic Elements in Architecture and Urban Design	Health and Well-Being Benefits
Visual Comfort	- Design for Natural Light - Quality and Quantity of Illumination (Natural and Artificial Light) - Access to Outdoor Views - Color and Form Harmony - Comfort Distance and Orientation	- Allowing Daylight and Balancing Artificial Light - Green Courtyards and Roofs - Vertical Green Walls and Façades - Nature Images and Paintings - Botanical Motifs - Green Streets and Pedestrian Connections	- Stress Reduction - Cognitive Enhancement - Improved Life Satisfaction
Air Quality	- Design for Natural Ventilation - Effective CO_2 Control - Integration of Plantable Resources - Ecosystem Balance - Clean Air Solutions	- Natural Airflow Distribution via Green Atriums - Indoor and Outdoor Plants - Pocket Parks and Public Landscaping - Low Impact Development (LID) - Sustainable Waste Management	- Better Air Quality - Respiratory Health Improvement - Enhanced Immune System - Humidity Control
Thermal Comfort	- Design for Natural Light and Ventilation - Selection of Thermal Mass Materials - Heat Exchange Surfaces - Water-absorbent Features - Vegetation Integration	- Allowing Natural Light and Ventilation - Thermally Treated Timber and Natural Materials like Clay, Wood, Stone - Vegetated Swales - Urban Parks and Trees - Water Features/Reservoirs	- Thermal Balance - Energy Efficiency - Reduced Heat Stress - Enhanced Comfort in Diverse Climates
Acoustics	- Sounds of Nature - Proper Sound Placement - Noise Control/ Sound Absorption - Decorative Acoustic Solutions	- Natural Soundscapes (Water, Wind, Wildlife) - Green Walls for Sound Absorption - Vegetative Buffers for Noise Reduction - Outdoor Quiet Zones - Green Streets and Roadside Parks	- Noise Reduction - Stress Relief - Acoustic Comfort - Improved Cognitive Function
Ergonomics	- Safe and Accessible Design - Human-Centered Spatial Planning - Productivity and Efficiency Focus - Aesthetic and Functional Integration	- Universal Design Principles - Human-centered Safety Measures - Gender and Age Considerations in Design - Flexible and Adaptable Spaces	- Enhanced Safety and Security - Increased Productivity - User Satisfaction - Sense of Belonging and Comfort

(1) Visual Comfort

Visual comfort emphasizes the importance of natural light, access to outdoor views, and the harmonious use of color and form [28,116]. The design incorporates elements such as green courtyards, vertical gardens, and nature-inspired aesthetics like botanical motifs, which together create visually stimulating and relaxing environments. These strategies reduce stress, enhance cognitive functions, and increase overall comfort [44,118,119]. By ensuring that users have access to nature, either directly or indirectly, through design, spaces become more inviting and reduce mental fatigue [31,43,120,121].

(2) Air Quality

Air quality is an essential aspect of human well-being, as poor air quality is associated with respiratory issues and overall discomfort [29,67,116]. This framework emphasizes natural ventilation, CO_2 control, and the integration of plants to purify the air. Biophilic design elements such as green atriums, pocket parks, and green roofs help filter pollutants, improve air circulation, and contribute to respiratory health and the immune system's strength. Enhanced air quality fosters a healthier and more productive environment in both residential and commercial spaces [74,107,108,122].

(3) Thermal Comfort

Thermal comfort is crucial for maintaining a balanced indoor environment [123,124]. This framework enhances thermal comfort by integrating natural materials, sustainable design practices, and features such as water bodies and vegetative cover. Elements such as vegetated swales, urban parks, and green walls help regulate temperature, reduce the urban heat island effect, and provide passive cooling, making the environment more comfortable in varying climates [12,44]. These strategies not only support human comfort but also improve energy efficiency and sustainability [3,87,125].

(4) Acoustics

Acoustic comfort focuses on minimizing noise pollution and enhancing natural soundscapes. The use of natural sounds, green walls, vegetative buffers, and quiet outdoor zones promotes a tranquil environment conducive to relaxation and cognitive function [126–128]. By reducing noise levels, biophilic design helps mitigate the stress associated with urban noise pollution, thereby improving occupant well-being [129,130].

(5) Ergonomics

Ergonomics in this biophilic framework ensures that environments are safe, accessible, and conducive to productivity. By applying universal design principles and integrating human-centered safety measures such as Crime Prevention Through Environmental Design (CPTED), spaces can cater to diverse needs while enhancing safety, accessibility, and satisfaction [63,131,132]. These considerations are essential in designing spaces that accommodate users' physical and psychological comfort, thus improving both functional and emotional experiences within built environments [37,133,134].

The newly developed biophilic design framework integrates research findings with innovative strategies to effectively target human comfort and satisfaction. By addressing visual, air quality, thermal, acoustic, and ergonomic factors, the framework offers a holistic approach to enhancing the well-being of occupants. Moreover, the inclusion of urban design elements broadens its applicability across various scales, from individual buildings to entire city blocks. As a result, this framework presents a flexible, comprehensive model for biophilic design that supports human health, comfort, and sustainable urban development.

5.3. Limitations and Gaps

While this framework is a significant advancement, it does have some minimal limitations:

(1) Outdoor Focus Could Be Enhanced: Although this table introduces a more balanced approach between indoor and outdoor design elements, the outdoor components could benefit from further development, particularly in urban contexts where outdoor space is limited. Future research could explore how to optimize outdoor biophilic interventions in dense urban areas where space constraints make it difficult to implement larger green infrastructure projects [116].

(2) Empirical Validation: While the performance metrics provided are detailed and measurable, further empirical validation through long-term studies will strengthen the framework. Existing frameworks often lack this kind of empirical backing, and while this table makes a step forward, its widespread adoption will require robust, data-driven evidence that demonstrates its effectiveness in various building types and climates.

This newly developed framework is more efficient in bridging the gap between biophilic design theory and practical application. It enhances the existing frameworks from Table 7 (a comparative analysis of biophilic design frameworks) by integrating specific human comfort and sustainability factors, offering measurable performance metrics and a structured approach for implementation. However, it also presents limitations, such as potential complexity, which might challenge designers in applying these concepts effectively. Overall, the integrated design in the Integrated Biophilic Design for Human Comfort and Sustainability framework addresses the shortcomings of earlier frameworks while emphasizing the need for a comprehensive and measurable approach to biophilic design.

Table 7. Comparative analysis of previously developed and newly developed biophilic design frameworks: fostering human comfort and sustainability.

Aspect	Existing—Biophilic Design Frameworks	Newly Developed—Integrated Biophilic Design for Human Comfort and Sustainability	Comparison
Main Focus	Connection between humans and nature; enhancing well-being.	Integration of various design dimensions for comfort and sustainability.	Integrated Biophilic Design builds on the foundational principles of the existing biophilic design frameworks but provides a more comprehensive and integrated approach that directly links comfort with sustainability goals.
Dimensions/Attributes	Attributes focus on environmental features and experiences with nature.	Incorporates multiple dimensions (e.g., direct and indirect experiences) with specific components related to comfort and sustainability.	Integrated Biophilic Design's detailed components, such as dynamic lighting and thermal variability, enhance the understanding of how biophilic design can be practically applied in architectural design.
Key Patterns/ Strategies	Emphasize types of experiences and strategies for sensory engagement.	Provides actionable strategies linked to specific performance metrics for each design dimension.	While the existing biophilic design frameworks outline broader strategies, Integrated Biophilic Design translates these strategies into measurable performance metrics, enhancing its practical application in design and evaluation.
Integration of IEQ and POE	Limited discussion on indoor environmental quality (IEQ) and post-occupancy evaluation (POE).	Explicitly integrates IEQ and POE for all components, focusing on user satisfaction and environmental quality.	Integrated Biophilic Design's emphasis on IEQ and POE makes it more suitable for contemporary design practices that prioritize user experience and environmental sustainability, addressing a gap in the existing biophilic design frameworks.
Performance Metrics	No performance metrics provided.	Specifies quantitative and qualitative metrics for evaluating comfort and sustainability.	Integrated Biophilic Design offers a robust framework for measuring success, which is crucial for validating biophilic design principles and their impact on human comfort and environmental performance.
Limitations	Provide a foundational understanding of biophilic design frameworks but lack specific applications.	While it includes comprehensive metrics and factors, it may become overly complex, making implementation challenging.	Integrated Biophilic Design, despite its strengths, risks overwhelming designers with too many components and metrics, which may hinder practical application in some contexts.

6. Conclusions

The comprehensive review of biophilic design frameworks highlights the significant potential of integrating natural elements into the built environment to enhance human comfort, satisfaction, and well-being. While existing frameworks, provide valuable insights into the psychological and physiological benefits of biophilic design, gaps remain in their application, particularly in urban contexts. These gaps include the need for clearer linkages between specific biophilic design elements and measurable health benefits, especially in dense urban areas where access to nature is limited.

This paper proposes an enhanced biophilic design framework that systematically categorizes and links biophilic elements to their corresponding psychological, physiological, and environmental benefits. By integrating both indoor and outdoor environmental factors, this framework offers a more holistic approach than previous models, ensuring a comprehensive application across varying scales—from individual buildings to broader urban settings. The framework emphasizes not only direct experiences of nature, such as natural light and vegetation, but also indirect experiences, such as views of nature and non-visual sensory stimuli, to foster deeper connections to nature.

The findings suggest that biophilic design contributes not only to environmental sustainability by improving energy efficiency and indoor air quality (IEQ) but also plays a crucial role in enhancing mental health, reducing stress, and fostering social well-being. This aligns with the growing recognition of biophilic design as a vital tool for creating resilient, healthy, and sustainable environments, particularly as urbanization continues to accelerate globally. The integration of outdoor urban elements, such as green roofs, biodiversity, and stormwater management, further extends the relevance of biophilic principles in achieving both human-centered and ecological design goals.

Future research should focus on refining and expanding this framework, particularly by exploring its application in diverse cultural and climatic contexts. Additionally, the development of standardized metrics for evaluating the long-term impacts of biophilic design on both human well-being and environmental health will be critical. Post-occupancy evaluation (POE) and user surveys will play a crucial role in validating the framework and ensuring its adaptability. This will facilitate the widespread adoption of biophilic design as a core component of sustainable urban development, ultimately improving the quality of life for urban inhabitants across the globe.

Author Contributions: Conceptualization, R.R.T. and J.P.; Data curation, R.R.T.; Formal analysis, R.R.T.; Methodology, R.R.T. and J.P.; Supervision, J.P.; Validation, J.P.; Writing—original draft, R.R.T.; Writing—review and editing, J.P. All authors have read and agreed to the published version of the manuscript.

Funding: This research received no external funding.

Data Availability Statement: The original contributions presented in the study are included in the article, further inquiries can be directed to the corresponding author.

Acknowledgments: The authors wish to express their sincere gratitude to the members of the Humans, Environment, and Architecture Lab (HEAL) at Ewha Womans University for their invaluable research guidance and support throughout the preparation of this paper.

Conflicts of Interest: The authors declare no conflicts of interest.

Abbreviations

IAQ	Indoor Air Quality
IEQ	Indoor Environmental Quality
POE	Post-occupancy Evaluation
ASHRAE	American Society of Heating, Refrigerating and Air-Conditioning Engineers

References

1. Kellert, S.R. *Building for Life: Designing and Understanding the Human-Nature Connection*; Island Press: Washington, DC, USA, 2012.
2. Kellert, S.R. Building for life: Designing and understanding the human-nature connection. *Renew. Resour. J.* **2006**, *24*, 8.
3. Wijesooriya, N.; Brambilla, A. Bridging biophilic design and environmentally sustainable design: A critical review. *J. Clean. Prod.* **2020**, *283*, 124591. [CrossRef]
4. Söderlund, J. *The Emergence of Biophilic Design*; Springer: Cham, Switzerland, 2019.
5. Zhong, W.; Schröder, T.; Bekkering, J. Biophilic design in architecture and its contributions to health, well-being, and sustainability: A critical review. *Front. Archit. Res.* **2022**, *11*, 114–141. [CrossRef]
6. Kellert, S.; Calabrese, E. *The Practice of Biophilic Design*; Terrapin Bright LLC: London, UK, 2015; Volume 3.
7. Yassein, G.; Ebrahiem, S. Biophilic design in the built environment to improve well-being: A systematic review of practices. *J. Urban Res.* **2018**, *30*, 128–146. [CrossRef]
8. Andreucci, M.B.; Loder, A.; Brown, M.; Brajković, J. Exploring challenges and opportunities of biophilic urban design: Evidence from research and experimentation. *Sustainability* **2021**, *13*, 4323. [CrossRef]
9. Bowler, D.E.; Buyung-Ali, L.M.; Knight, T.M.; Pullin, A.S. A systematic review of evidence for the added benefits to health of exposure to natural environments. *BMC Public Health* **2010**, *10*, 456. [CrossRef]
10. Kabinesh, V.; Vennila, S.; Baranidharan, K.; Ravi, R.; Krishnamoorthi, S.; Thirunavukkarasu, M. Sustainable Spaces-The Evolution of Biophilic Design in Modern Architecture: A Review. *Asian J. Environ. Ecol.* **2024**, *23*, 64–77. [CrossRef]
11. Kellert, S.R.; Heerwagen, J.; Mador, M. *Biophilic Design: The Theory, Science and Practice of Bringing Buildings to Life*; John Wiley & Sons: Hoboken, NJ, USA, 2011.
12. Africa, J.; Heerwagen, J.; Loftness, V.; Balagtas, C.R. Biophilic design and climate change: Performance parameters for health. *Front. Built Environ.* **2019**, *5*, 28. [CrossRef]
13. Kaplan, S. The restorative benefits of nature: Toward an integrative framework. *J. Environ. Psychol.* **1995**, *15*, 169–182. [CrossRef]
14. Kaplan, R.; Kaplan, S. *The Experience of Nature: A Psychological Perspective*; Cambridge University Press: Cambridge, UK, 1989.
15. Beatley, T. *Biophilic Cities: Integrating Nature into Urban Design and Planning*; Island Press: Washington, DC, USA, 2011.
16. Calabrese, E.F.; Dommert, A. Biophilia and the Practice of Biophilic Design. In *Pathways to Well-Being in Design*; Routledge: London, UK, 2018; pp. 97–127.
17. Browning, W.D.; Ryan, C.O. *Nature Inside: A Biophilic Design Guide*; Routledge: London, UK, 2020.
18. Ryan, C.O.; Browning, W.D. Biophilic design. In *Sustainable Built Environments*; Springer: New York, NY, USA, 2020; pp. 43–85.
19. Daroda, K.S. Climate responsive architecture: Creating greater design awareness among architects. *J. Environ. Issues Agric. Dev. Ctries.* **2011**, *3*, 33.
20. Sinha, A. Building orientation as the primary design consideration for climate responsive architecture in urban areas. *Archit. Urban Plan.* **2020**, *16*, 32–40. [CrossRef]
21. Ragheb, A.; El-Shimy, H.; Ragheb, G. Green architecture: A concept of sustainability. *Procedia-Soc. Behav. Sci.* **2016**, *216*, 778–787. [CrossRef]
22. Aliamin, Y. Pathways toward sustainable architecture: Green architecture and circular built environment. In *IOP Conference Series: Earth and Environmental Science*; IOP Publishing: Bristol, UK, 2021.
23. Bielek, B. Green building–towards sustainable architecture. *Appl. Mech. Mater.* **2016**, *824*, 751–760. [CrossRef]
24. Guzowski, M. Biophilic Net-Positive Architecture: Integrating Nature, Health, Wellbeing and Passive Design. In Proceedings of the Architectural Research Centers Consortium, Virtual, 15–18 June 2016; Volume 21.
25. Hootman, T. *Net Zero Energy Design: A Guide for Commercial Architecture*; John Wiley & Sons: Hoboken, NJ, USA, 2012.
26. Downton, P.; Jones, D.; Zeunert, J.; Roös, P. Biophilic design applications: Putting theory and patterns into built environment practice. *KnE Eng.* **2017**, *2*, 59–65. [CrossRef]
27. Lei, Q.; Lau, S.S.Y.; Yuan, C.; Qi, Y. Post-occupancy evaluation of the biophilic design in the workplace for health and wellbeing. *Buildings* **2022**, *12*, 417. [CrossRef]
28. Park, J.; Loftness, V.; Aziz, A.; Wang, T.-H. Strategies to achieve optimum visual quality for maximum occupant satisfaction: Field study findings in office buildings. *Build. Environ.* **2021**, *195*, 107458. [CrossRef]
29. Park, J.; Loftness, V.; Aziz, A.; Wang, T.H. Critical factors and thresholds for user satisfaction on air quality in office environments. *Build. Environ.* **2019**, *164*, 106310. [CrossRef]
30. Ryan, M.M. (Ed.) *Handbook of US Labor Statistics 2015: Employment, Earnings, Prices, Productivity, and Other Labor Data*; Bernan Press: Lanham, MD, USA.
31. Hung, S.-H.; Chang, C.-Y. Health benefits of evidence-based biophilic-designed environments: A review. *J. People Plants Environ.* **2021**, *24*, 1–16. [CrossRef]
32. Heerwagen, J. Biophilia, health, and well-being. In *Restorative Commons: Creating Health and Well-Being through Urban Landscapes*; Government Printing Office: Washington, DC, USA, 2009; 39, pp. 39–57.
33. Bolten, B.; Barbiero, G. Biophilic Design: How to Enhance Physical and Psychological Health and Wellbeing in Our Built Environments. 2020. Available online: https://ojs.unito.it/index.php/visions/article/view/3829 (accessed on 8 September 2024).
34. Staszowska, A. Application of Biophilic Installations for Indoor Air Quality Improvement. *Rocz. Ochr. Sr.* **2020**, *22*, 716–726.
35. Al-Dmour, Y. *Flourishing the Indoor Environment Quality of Workplaces Using the Biophilic Architectural Design*; Brunel University London: London, UK, 2023.

36. Kavathekar, A.; Bantanur, S. Study of the influence of indoor plants as an indicator of biophilic design on CO2 concentrations in a classroom of higher education institute. *J. Sustain. Archit. Civ. Eng.* **2022**, *31*, 96–108. [CrossRef]
37. Lei, Q.; Yuan, C.; Lau, S.S.Y. A quantitative study for indoor workplace biophilic design to improve health and productivity performance. *J. Clean. Prod.* **2021**, *324*, 129168. [CrossRef]
38. Dalay, L. The impact of biophilic design elements on the atmospheric perception of the interior space. *Uluslararası Peyzaj Mimarlığı Araştırmaları Dergisi (IJLAR)* **2020**, *4*, 4–20.
39. Ryan, C.O.; Browning, W.D.; Clancy, J.O.; Andrews, S.L.; Kallianpurkar, N.B. Biophilic design patterns: Emerging nature-based parameters for health and well-being in the built environment. *ArchNet-IJAR Int. J. Archit. Res.* **2014**, *8*, 62. [CrossRef]
40. Richardson, M.; Butler, C.W. Nature connectedness and biophilic design. *Build. Res. Inf.* **2022**, *50*, 36–42. [CrossRef]
41. Grinde, B.; Patil, G.G. Biophilia: Does visual contact with nature impact on health and well-being? *Int. J. Environ. Res. Public Health* **2009**, *6*, 2332–2343. [CrossRef] [PubMed]
42. Gillis, K.; Gatersleben, B. A review of psychological literature on the health and wellbeing benefits of biophilic design. *Buildings* **2015**, *5*, 948–963. [CrossRef]
43. Berto, R.; Barbiero, G.; Pasini, M.; Pieter, U. Biophilic design triggers fascination and enhances psychological restoration in the urban environment. *J. Biourbanism* **2015**, *1*, 27–34.
44. Yin, J.; Zhu, S.; MacNaughton, P.; Allen, J.G.; Spengler, J.D. Physiological and cognitive performance of exposure to biophilic indoor environment. *Build. Environ.* **2018**, *132*, 255–262. [CrossRef]
45. Edwards, B. *Sustainable Architecture: European Directives and Building Design*; Architectual Press: New York, NY, USA, 1999.
46. Makram, A.; Abou Ouf, T. Biomimetic and biophilic design as an approach to innovative sustainable architectural design. In Proceedings of the AR-UP 2019: Third International Conference of Architecture and Urban Planning, Cairo, Egypt, 24–26 October 2019; Department Faculty of Engineering, Ain Shams University,: Cairo, Egypt.
47. Jaheen, N.U.; El-Darwish, I. Biophilic Design Elements in Modern Buildings Influenced by Islamic Architecture Features. *JES J. Eng. Sci.* **2022**, *50*, 41–54. [CrossRef]
48. Abdelaal, M.; Soebarto, V. History matters: The origins of biophilic design of innovative learning spaces in traditional architecture. *ArchNet—IJAR* **2018**, *12*, 108–127. [CrossRef]
49. Dalay, L.; Aytaç, G. Biophilic design: Integrating nature into the urban environment. In *Emerging Approaches in Design and New Connections with Nature*; IGI Global: Hershey, PA, USA, 2022; pp. 1–19.
50. Kellert, S.R.; Wilson, E.O. *The Biophilia Hypothesis*; Island Press: Washington, DC, USA, 1995.
51. Lefosse, D.; van Timmeren, A.; Ratti, C. Biophilia upscaling: A systematic literature review based on a three-metric approach. *Sustainability* **2023**, *15*, 15702. [CrossRef]
52. Kellert, S.R. *Nature by Design: The Practice of Biophilic Design*; Yale University Press: New Haven, CT, USA, 2018.
53. Barbiero, G.; Berto, R. Biophilia as evolutionary adaptation: An onto-and phylogenetic framework for biophilic design. *Front. Psychol.* **2021**, *12*, 700709. [CrossRef] [PubMed]
54. Biederman, I.; Vessel, E.A. Perceptual pleasure and the brain: A novel theory explains why the brain craves information and seeks it through the senses. *Am. Sci.* **2006**, *94*, 247–253. [CrossRef]
55. Brielmann, A.A.; Buras, N.H.; Salingaros, N.A.; Taylor, R.P. What happens in your brain when you walk down the street? implications of architectural proportions, biophilia, and fractal geometry for urban science. *Urban Sci.* **2022**, *6*, 3. [CrossRef]
56. Valentine, C.; Steffert, T.; Mitcheltree, H.; Steemers, K. Architectural Neuroimmunology: A Pilot Study Examining the Impact of Biophilic Architectural Design on Neuroinflammation. *Buildings* **2024**, *14*, 1292. [CrossRef]
57. Hartig, T.; Berg, A.E.V.D.; Hagerhall, C.M.; Tomalak, M.; Bauer, N.; Hansmann, R.; Ojala, A.; Syngollitou, E.; Carrus, G.; van Herzele, A.; et al. Health benefits of nature experience: Psychological, social and cultural processes. In *Forests, Trees and Human Health*; Springer: Dordrecht, The Netherlands, 2011; pp. 127–168.
58. Wyles, K.J.; White, M.P.; Hattam, C.; Pahl, S.; King, H.; Austen, M. Are some natural environments more psychologically beneficial than others? The importance of type and quality on connectedness to nature and psychological restoration. *Environ. Behav.* **2019**, *51*, 111–143. [CrossRef]
59. Ward Thompson, C.; Aspinall, P.A. Natural environments and their impact on activity, health, and quality of life. *Appl. Psychol. Health Well-Being* **2011**, *3*, 230–260. [CrossRef]
60. Hartig, T.; Evans, G.W. Psychological foundations of nature experience. In *Advances in Psychology*; Elsevier: Amsterdam, The Netherlands, 1993; pp. 427–457.
61. Thompson Coon, J.; Boddy, K.; Stein, K.; Whear, R.; Barton, J.; Depledge, M.H. Does participating in physical activity in outdoor natural environments have a greater effect on physical and mental wellbeing than physical activity indoors? A systematic review. *Environ. Sci. Technol.* **2011**, *45*, 1761–1772. [CrossRef]
62. Tekin, B.H.; Corcoran, R.; Gutiérrez, R.U. A systematic review and conceptual framework of biophilic design parameters in clinical environments. *HERD Health Environ. Res. Des. J.* **2023**, *16*, 233–250. [CrossRef]
63. Soderlund, J.; Newman, P. Biophilic architecture: A review of the rationale and outcomes. *AIMS Environ. Sci.* **2015**, *2*, 950–969.
64. Buceta-Albillos, N.; Ayuga-Téllez, E. The State of the Art in the Biophilic Construction of Healthy Spaces for People. *Buildings* **2024**, *14*, 491. [CrossRef]
65. Makes, N.I.M.; Sense, F. *The Economics of Biophilia*; Terrapin Bright Green, LLC: New York, NY, USA, 2012.

66. Untaru, E.-N.; Han, H.; David, A.; Chi, X. Biophilic design and its effectiveness in creating emotional well-being, green satisfaction, and workplace attachment among healthcare professionals: The hospice context. *HERD Health Environ. Res. Des. J.* **2023**, *17*, 190–208. [CrossRef] [PubMed]
67. Loftness, V.; Aziz, A.; Choi, J.; Kampschroer, K.; Powell, K.; Atkinson, M.; Heerwagen, J. The value of post-occupancy evaluation for building occupants and facility managers. *Intell. Build. Int.* **2009**, *1*, 249–268. [CrossRef]
68. Zijlema, W.L.; Triguero-Mas, M.; Smith, G.; Cirach, M.; Martinez, D.; Dadvand, P.; Gascon, M.; Jones, M.; Gidlow, C.; Hurst, G.; et al. The relationship between natural outdoor environments and cognitive functioning and its mediators. *Environ. Res.* **2017**, *155*, 268–275. [CrossRef]
69. Johnson, P. Growing physical, social and cognitive capacity: Engaging with natural environments. *Int. Educ. J.* **2007**, *8*, 293–303.
70. Berman, M.G.; Jonides, J.; Kaplan, S. The cognitive benefits of interacting with nature. *Psychol. Sci.* **2008**, *19*, 1207–1212. [CrossRef]
71. Schertz, K.E.; Berman, M.G. Understanding nature and its cognitive benefits. *Curr. Dir. Psychol. Sci.* **2019**, *28*, 496–502. [CrossRef]
72. Bailey, A.W.; Allen, G.; Herndon, J.; Demastus, C. Cognitive benefits of walking in natural versus built environments. *World Leis. J.* **2018**, *60*, 293–305. [CrossRef]
73. Gullone, E. The biophilia hypothesis and life in the 21st century: Increasing mental health or increasing pathology? *J. Happiness Stud.* **2000**, *1*, 293–322. [CrossRef]
74. Moslehian, A.S.; Roös, P.B.; Gaekwad, J.S.; Van Galen, L. Potential risks and beneficial impacts of using indoor plants in the biophilic design of healthcare facilities: A scoping review. *Build. Environ.* **2023**, *233*, 110057. [CrossRef]
75. Albert, C.; Schröter, B.; Haase, D.; Brillinger, M.; Henze, J.; Herrmann, S.; Gottwald, S.; Guerrero, P.; Nicolas, C.; Matzdorf, B. Addressing societal challenges through nature-based solutions: How can landscape planning and governance research contribute? *Landsc. Urban Plan.* **2019**, *182*, 12–21. [CrossRef]
76. Hähn, N.; Essah, E.; Blanusa, T. Biophilic design and office planting: A case study of effects on perceived health, well-being and performance metrics in the workplace. *Intell. Build. Int.* **2021**, *13*, 241–260. [CrossRef]
77. Chen, Y. *The Impact of Biophilic Design on Health and Wellbeing of Residents through Raising Environmental Awareness and Nature Connectedness*; University of Georgia Athens: Athens, GA, USA, 2017.
78. Hung, S.-H.; Chang, C.-Y. How do humans value urban nature? Developing the perceived biophilic design scale (PBDs) for preference and emotion. *Urban For. Urban Green.* **2022**, *76*, 127730. [CrossRef]
79. Reeve, A.C.; Desha, C.; Hargreaves, D.; Hargroves, K. Biophilic urbanism: Contributions to holistic urban greening for urban renewal. *Smart Sustain. Built Environ.* **2015**, *4*, 215–233. [CrossRef]
80. Nejade, R.M.; Grace, D.; Bowman, L.R. What is the impact of nature on human health? A scoping review of the literature. *J. Glob. Health* **2022**, *12*, 04099. [CrossRef] [PubMed]
81. Yildirim, M.; Gocer, O.; Globa, A.; Brambilla, A. Investigating restorative effects of biophilic design in workplaces: A systematic review. *Intell. Build. Int.* **2023**, *15*, 205–247. [CrossRef]
82. Gray, T. Retrofitting biophilic design elements into office site sheds: Does' going green'enhance the well-being and productivity of workers. In *Landscape Architecture: The Sense of Places, Models and Applications*; Jönköping University: Jönköping, Sweden, 2017; pp. 105–126.
83. Topgül, S. The impact of biophilic designs on worker efficiency. *Sos. Araştırmalar Davranış Bilim.* **2019**, *5*, 11–26.
84. Novosadová, L.; van der Knaap, W. The role of biophilic agents in building a green resilient city; the case of Birmingham, UK. *Sustainability* **2021**, *13*, 5033. [CrossRef]
85. Parsaee, M.; Demers, C.M.; Hébert, M.; Lalonde, J.-F.; Potvin, A. A photobiological approach to biophilic design in extreme climates. *Build. Environ.* **2019**, *154*, 211–226. [CrossRef]
86. Zhong, W.; Schröder, T.; Bekkering, J. Implementing biophilic design in architecture through three-dimensional green spaces: Guidelines for building technologies, plant selection, and maintenance. *J. Build. Eng.* **2024**, *92*, 109648. [CrossRef]
87. Parsaee, M.; Demers, C.M.; Hébert, M.; Lalonde, J.-F.; Potvin, A. Biophilic, photobiological and energy-efficient design framework of adaptive building façades for Northern Canada. *Indoor Built Environ.* **2021**, *30*, 665–691. [CrossRef]
88. Sadick, A.-M.; Kamardeen, I.; Vu, X.P. Challenges for implementing biophilic strategies in Australian building design. *J. Build. Eng.* **2023**, *74*, 106849. [CrossRef]
89. Gill, S.E.; Handley, J.F.; Ennos, A.R.; Pauleit, S. Adapting cities for climate change: The role of the green infrastructure. *Built Environ.* **2007**, *33*, 115–133. [CrossRef]
90. Bush, J.; Doyon, A. Building urban resilience with nature-based solutions: How can urban planning contribute? *Cities* **2019**, *95*, 102483. [CrossRef]
91. Isimbi, D.; Park, J. The Analysis of the EDGE Certification System on Residential Complexes to Improve Sustainability and Affordability. *Buildings* **2022**, *12*, 1729. [CrossRef]
92. Beatley, T. Biophilic urbanism: Inviting nature back to our communities and into our lives. *Wm. Mary Envtl. L. Pol'y Rev.* **2009**, *34*, 209.
93. Lee, E.J.; Park, S.J. Toward the biophilic residential regeneration for the green new deal. *Int. J. Environ. Res. Public Health* **2021**, *18*, 2523. [CrossRef]
94. Browning, W. *Constructing the Biophilic Community*; The MIT Press: Cambridge, MA, USA, 2013.
95. Threlfall, C.G.; Mata, L.; Mackie, J.A.; Hahs, A.K.; Stork, N.E.; Williams, N.S.G.; Livesley, S.J. Increasing biodiversity in urban green spaces through simple vegetation interventions. *J. Appl. Ecol.* **2017**, *54*, 1874–1883. [CrossRef]

96. Aronson, M.F.; Lepczyk, C.A.; Evans, K.L.; Goddard, M.A.; Lerman, S.B.; MacIvor, J.S.; Nilon, C.H.; Vargo, T. Biodiversity in the city: Key challenges for urban green space management. *Front. Ecol. Environ.* **2017**, *15*, 189–196.
97. Hu, X.; Lima, M.F. The association between maintenance and biodiversity in urban green spaces: A review. *Landsc. Urban Plan.* **2024**, *251*, 105153. [CrossRef]
98. Sadler, J.; Bates, A.; Hale, J.; James, P. Bringing cities alive: The importance of urban green spaces for people and biodiversity. In *Urban Ecology*; Cambridge University Press: Cambridge, UK, 2010; pp. 230–260.
99. Fukumoto, H.; Shimoda, M.; Hoshino, S. The effects of different designs of indoor biophilic greening on psychological and physiological responses and cognitive performance of office workers. *PLoS ONE* **2024**, *19*, e0307934.
100. Tharim, A.H.A.; Ahmad, A.C.; Saarani, P.S.N.; Haossain, N.Q.R.; Purwanto, E.; Tafridj, I.S.I.; Purisari, R.; Prasetio, T. The determinant factors of biophilic design strategies and occupants'psychological performance in office building. *Malays. J. Sustain. Environ.* **2022**, *9*, 87–106.
101. Pedersen Zari, M. Understanding and designing nature experiences in cities: A framework for biophilic urbanism. *Cities Health* **2023**, *7*, 201–212. [CrossRef]
102. Lin, B.B.; Egerer, M.H.; Ossola, A. Urban gardens as a space to engender biophilia: Evidence and ways forward. *Front. Built Environ.* **2018**, *4*, 79. [CrossRef]
103. Totaforti, S. Emerging biophilic urbanism: The value of the human–nature relationship in the urban space. *Sustainability* **2020**, *12*, 5487. [CrossRef]
104. Klotz, A.C.; Bolino, M.C. Bringing the great outdoors into the workplace: The energizing effect of biophilic work design. *Acad. Manag. Rev.* **2021**, *46*, 231–251. [CrossRef]
105. Barbiero, G.; Berto, R.; Venturella, A.; Maculan, N. Bracing Biophilia: When biophilic design promotes pupil's attentional performance, perceived restorativeness and affiliation with Nature. *Environ. Dev. Sustain.* **2021**, 1–15. [CrossRef]
106. Zhong, W.; Schroeder, T.; Bekkering, J. Designing with nature: Advancing three-dimensional green spaces in architecture through frameworks for biophilic design and sustainability. *Front. Archit. Res.* **2023**, *12*, 732–753. [CrossRef]
107. Yeo, N.L.; Elliott, L.R.; Bethel, A.; White, M.P.; Dean, S.G.; Garside, R. Indoor nature interventions for health and wellbeing of older adults in residential settings: A systematic review. *Gerontologist* **2020**, *60*, e184–e199. [CrossRef]
108. Thatcher, A.; Adamson, K.; Bloch, L.; Kalantzis, A. Do indoor plants improve performance and well-being in offices? Divergent results from laboratory and field studies. *J. Environ. Psychol.* **2020**, *71*, 101487. [CrossRef]
109. Han, K.-T.; Ruan, L.-W. Effects of indoor plants on self-reported perceptions: A systemic review. *Sustainability* **2019**, *11*, 4506. [CrossRef]
110. Bringslimark, T.; Hartig, T.; Patil, G.G. The psychological benefits of indoor plants: A critical review of the experimental literature. *J. Environ. Psychol.* **2009**, *29*, 422–433. [CrossRef]
111. Larsen, L.; Adams, J.; Deal, B.; Kweon, B.S.; Tyler, E. Plants in the workplace: The effects of plant density on productivity, attitudes, and perceptions. *Environ. Behav.* **1998**, *30*, 261–281. [CrossRef]
112. Huntsman, D.D.; Bulaj, G. Healthy dwelling: Design of biophilic interior environments fostering self-care practices for people living with migraines, chronic pain, and depression. *Int. J. Environ. Res. Public Health* **2022**, *19*, 2248. [CrossRef] [PubMed]
113. Dodman, D.; Bicknell, J.; Satterthwaite, D. *Adapting Cities to Climate Change: Understanding and Addressing the Development Challenges*; Routledge: London, UK, 2012.
114. Ramyar, R.; Ackerman, A.; Johnston, D.M. Adapting cities for climate change through urban green infrastructure planning. *Cities* **2021**, *117*, 103316. [CrossRef]
115. Watchman, M.; Demers, C.M.; Potvin, A. Biophilic school architecture in cold climates. *Indoor Built Environ.* **2021**, *30*, 585–605. [CrossRef]
116. Loftness, V.; Hartkopf, V.; Aziz, A.; Choi, J.H.; Park, J. Critical Frameworks for Building Evaluation: User Satisfaction, Environmental Measurements and the Technical Attributes of Building Systems (POE + M). In *Building Performance Evaluation: From Delivery Process to Life Cycle Phases*; Preiser, W.F.E., Hardy, A.E., Schramm, U., Eds.; Springer International Publishing: Cham, Switzerland, 2018; pp. 29–48.
117. Park, J.; Aziz, A.; Loftness, V. Post-occupancy evaluation for energy conservation, superior IEQ & increased occupant satisfaction. In *IFMA's World Workplace 2013*; Ifma: 2013; International Facility Management Association: Pennsylvania, PA, USA, 2013.
118. Yin, J.; Yuan, J.; Arfaei, N.; Catalano, P.J.; Allen, J.G.; Spengler, J.D. Effects of biophilic indoor environment on stress and anxiety recovery: A between-subjects experiment in virtual reality. *Environ. Int.* **2020**, *136*, 105427. [CrossRef]
119. Gaekwad, J.S.; Moslehian, A.S.; Roös, P.B. A meta-analysis of physiological stress responses to natural environments: Biophilia and Stress Recovery Theory perspectives. *J. Environ. Psychol.* **2023**, *90*, 102085. [CrossRef]
120. Loftness, V.; Aziz, A.; Lam, K.P.; Lee, S.; Cochran, E.; Park, J.; Srivastava, R. *State of the Art in Enclosure Technologies and Integrated Systems for 50% Energy Savings in Existing Commercial Buildings*; REPORT-CBEI: Pittsburgh, PA, USA, 2012.
121. Aristizabal, S.; Byun, K.; Porter, P.; Clements, N.; Campanella, C.; Li, L.; Mullan, A.; Ly, S.; Senerat, A.; Nenadic, I.Z.; et al. Biophilic office design: Exploring the impact of a multisensory approach on human well-being. *J. Environ. Psychol.* **2021**, *77*, 101682. [CrossRef]
122. Schweighart, R.; Bettinger, L. Psychological and physiological effects of visual stimulation with artificial natural structures and indoor plants: A randomized controlled trial. *J. Urban Des. Ment. Health* **2023**, *8*, 513–522.
123. Park, J.; Loftness, V.; Aziz, A. Post-Occupancy Evaluation and IEQ Measurements from 64 Office Buildings: Critical Factors and Thresholds for User Satisfaction on Thermal Quality. *Buildings* **2018**, *8*, 156. [CrossRef]

124. Tham, K.W.; Willem, H.C. Room air temperature affects occupants' physiology, perceptions and mental alertness. *Build. Environ.* **2010**, *45*, 40–44. [CrossRef]
125. Nitu, M.A.; Gocer, O.; Wijesooriya, N.; Vijapur, D.; Candido, C. A biophilic design approach for improved energy performance in retrofitting residential projects. *Sustainability* **2022**, *14*, 3776. [CrossRef]
126. Newsham, G.; Birt, B.; Arsenault, C.; Thompson, L.; Veitch, J.; Mancini, S.; Galasiu, A.; Gover, B.; Macdonald, I.; Burns, G. Do green buildings outperform conventional buildings? Indoor environment and energy performance in North American offices. In *Research Report (National Research Council of Canada. Construction)*; no. RR-329; National Research Council of Canada: Ottawa, ON, Canada, 2012.
127. Park, J. Are Humans Good Sensors?: Using Occupants as Sensors for Indoor Environmental Quality Assessment and for Developing Thresholds that Matte. In *Architecture*; Carnegie Mellon University: Pittsburgh, PA, USA, 2015.
128. Park, J.; Loftness, V.; Wang, T.-H. Examining In Situ Acoustic Conditions for Enhanced Occupant Satisfaction in Contemporary Offices. *Buildings* **2022**, *12*, 1305. [CrossRef]
129. Pallasmaa, J. *The Eyes of the Skin: Architecture and the Senses*; John Wiley & Sons: Hoboken, NJ, USA, 2024.
130. Hadjiphilippou, P. *The Contribution of the Five Human Senses towards the Perception of Space*; Department of Architecture, University of Nicosia: Nicosia, Cyprus, 2013.
131. Söderlund, J.; Newman, P. Improving mental health in prisons through biophilic design. *Prison. J.* **2017**, *97*, 750–772. [CrossRef]
132. Shepley, M.; Sachs, N.; Sadatsafavi, H.; Fournier, C.; Peditto, K. The impact of green space on violent crime in urban environments: An evidence synthesis. *Int. J. Environ. Res. Public Health* **2019**, *16*, 5119. [CrossRef] [PubMed]
133. Sanchez, J.A.; Ikaga, T.; Sanchez, S.V. Quantitative improvement in workplace performance through biophilic design: A pilot experiment case study. *Energy Build.* **2018**, *177*, 316–328. [CrossRef]
134. Inglese, G.; Mura, A.L.; Alves, S.; Bonaiuto, M.; Villani, T. Biophilic design for remote studying environments: Analysis of case studies involving a collaboration between ergonomics and environmental psychology. *Connectivity* **2023**, *98*.

buildings

MDPI

Article

Construction and Optimization Strategies for Rural Residential Spatial Models Based on the Concept of Resource Metabolism: A Case Study of Rural Areas in the Shandong Plain

Zhongkai Zhou [1,2], Meng Liu [3,*], Jilong Zhao [1] and Chaoxian Li [1,*]

1 School of Architecture and Urban Planning, Shandong Jianzhu University, Jinan 250101, China
2 School of Architecture, Tianjin University, Tianjin 300072, China
3 School of Architecture and Engineering, Qingdao Binhai University, Qingdao 266555, China
* Correspondence: 06843@qdbhu.edu.cn (M.L.); 13946@sdjzu.edu.cn (C.L.)

Abstract: Rural communities can be conceptualized as spatial organisms interconnected by optimized resource utilization systems. Investigating the efficient utilization of rural resources and spatial construction methods grounded in resource metabolism is a pivotal step toward achieving the ecological transformation of rural spaces. This paper examined rural settlements in the Shandong Plain as a case study, exploring the relationships among three scales: village, neighborhoods, and courtyards. This analysis was based on elucidating the interaction mechanisms between "space and resource" and the integration of key resources and spatial elements. From the perspective of resource circulation and metabolism, this study aimed to elucidate the equilibrium of each resource element within three resource metabolism subsystems: the agricultural production system (core element), the ecological technology system (technological link), and the human life system (spatial carrier) in Shandong Plain's villages, considering general climatic conditions. To achieve this, this research utilized the resource production volume, the utilization and transformation volume of resource metabolism technology facilities per unit area, and the average per capita resource consumption as fundamental measurement units. The concept of a rationing relationship is introduced to clarify resource allocation. Combining the aforementioned research on spatial resource metabolism in ecological villages in Shandong Province with the material flow analysis method, this study constructed a bottom–up spatial model of resource metabolism at three scales, courtyards, neighborhoods, and villages, under various resource metabolism scenarios. This study is anticipated to significantly contribute to the theoretical understanding of rural habitat environments, offering novel methods and perspectives for constructing ecological rural settlements.

Keywords: resource metabolism; Shandong Plain-Type Rural Area; spatial modelling; Optimal Design Strategy; material flow analysis

1. Introduction

Exploring efficient and sustainable patterns of resource use and spatial construction methods is a significant avenue for alleviating urban–rural conflicts and achieving the transformation and development of rural areas. The current approach to resource utilization in China's rural areas is characterized by imprecision, hindering the rational allocation and efficient use of resources. Consequently, this increases in both per capita and overall resource consumption in rural areas poses significant challenges to their resource carrying capacity and energy transformation. Furthermore, the integration of rural production, living styles, and spatial elements is deficient, resulting in uniform spatial patterns of rural settlements and a trend toward "urbanization". As a major agricultural province in northern China, Shandong boasts diverse rural habitats and natural resources. Its sustainable remediation strategy and development model for rural settlements in the plains, the most prevalent rural settlement type in the province, has long served as a valuable reference and exemplar for similar initiatives across China. Therefore, it is crucial to consider the potential positive impact on enhancing the quality of the rural habitat environment in China, particularly in the northern region. However, a key challenge remains in achieving the ecological transformation of rural settlement spaces, based on understanding the intrinsic flow laws of rural resource elements and the characteristics of suitable resource utilization technologies. Nevertheless, there is a dearth of systematic theoretical research and practical investigation into achieving the ecological transformation and rational construction of rural settlements, based on the intrinsic flow laws of rural resource elements and the technical characteristics of suitable resource utilization.

Rural settlements are the most important spatial carriers for the lives of rural residents, and they are also the common object of research in a number of disciplines. In the early stages of research in related fields, fewer studies crossed disciplinary boundaries to pay in-depth attention to the construction of rural human settlements. In recent years, based on the cross-disciplinary perspective, taking the efficient utilization of resources as the goal and the quantitative analysis of resource flow as the main method, we have devoted ourselves to exploring the theory of "resource metabolism", which integrates the efficient utilization of resources and human settlements, opening up a brand-new perspective of systematically examining the coupled and coordinated relationship between the human settlement environment and the natural environment. It has become an important theory for solving the problem of sustainable development in rural settlements. It has opened up a new perspective for systematically scrutinizing the coupled and coordinated relationship between human settlements and the natural environment, and it has become an important theory and frontier perspective for solving the sustainable development of rural settlements. In terms of theoretical research on the connection between human habitat space and resource metabolism, the American ecologist Abel Wolman put forward the concept of "Urban Metabolism", which combines the concept of resource metabolism and related methods with the study of urban human habitat space [1]. Since then, relevant scholars and institutions have extended the concept of resource metabolism to the urban economy and industrial system, studying the metabolic flow process of various elements such as capital, raw material, water, waste, and other factors at different scales within the boundaries of urban and rural systems, as well as their impact assessment on the environment [2–5]. Agudelo-Vera pointed out that the cyclic efficiency and balanced management of resource metabolism are key factors in optimizing the sustainable development of urban and rural human settlements [6] and that recycling and the efficient use of resources reduces the dependence and impact of human settlements on external resources [7]. Based on the concept of resource metabolism, Kennedy has established indicators for the sustainable development of built spaces, which can be used to account for greenhouse gas emissions, construct dynamic analytical models, and set up sustainable design tools to dynamically evaluate the sustainable development of human settlements [8]. Researchers at the University of Groningen proposed the FEW-Nexus design tool and methodology based on the concept of resource metabolism from the perspective of the interaction between food, energy, and water resources, which affects the function of the human environment system. They furthermore carried out experimental design explorations on objects at different scales, such as neighborhoods, blocks, and districts, in some areas of the Netherlands and Japan [9]. The Dutch design office "upuse Studio", together with a team of researchers (REAP), proposed the "Cyclifiler" design strategy, which takes into account the characteristics of the transformation, distribution, and consumption processes of various types of resources, in an attempt to construct a spatial operation model that was compatible with the metabolism of resources and that advanced layer by layer from the bottom to the top [10,11]. Based on the theory of urban metabolism, Zhao Qianjun explored the driving relationship, optimization path, and evaluation mechanism of the metabolic process and efficiency on the functional layout and spatial form of urban settlements from a cross-disciplinary perspective [12].

From the systematic view of resource metabolism, the space of rural settlements can be regarded as a metabolic organism, which is in essence an ecosystem of various resource and energy flows. Incorporating metabolic thinking implies the synergistic integration of resource management and sustainable design of rural space, which in essence explores the way of efficient utilization of various resources in the countryside under the support of an appropriate spatial system. Metabolic research is therefore more direct and closer to the essence. Metabolism research, which focuses on the flow process and efficiency of resource elements, is more direct and close to the essence perspective method, which can reveal the dynamic change mechanism of rural settlements from a deeper level. The eco-village constructions developed in Europe and the United States since the 1970s can be regarded as the initial practice of combining the rural habitat with resource metabolism, and their distinctive feature is that a large number of cash crops have been planted in the countryside to act as greening, which is organically integrated with passive solar energy, a good walking environment, and natural drainage to form a complete ecosystem [13]. With the popularization and application of eco-agriculture in China's rural areas in the 1980s, Chinese scholars in the field of agricultural economics paid early attention to the energy flow and material cycle of rural ecosystems, laying a theoretical foundation for the application of resource metabolism analysis to the study of rural spatial problems [14]. For example, Wombach pointed out that eco-villages should make full use of natural resources within their spatial boundaries, accelerating material circulation and energy transformation, so as to realize simultaneous improvement in eco-agriculture and the living environment of human beings [15]. Some scholars have summarized the four models, construction principles, and technical components of eco-village construction in China [16], or the construction of an eco-village resource metabolism technology system containing energy, nutrient, and water cycles based on the principles of ecology, with agricultural production as the core [17]. The theory of resource metabolism has been applied to the study of rural human settlement space, but so far, there is a lack of systematic quantitative research methodology and effective application at the level of rural settlements with small scales.

Based on the above considerations, this study used resource cycle metabolism as an entry point. By integrating theoretical studies on resource metabolism with practical cases of urban and rural settlements, it explores the metabolic flow processes of resource elements such as water, energy, food, and waste in rural settlements in the plain area of Shandong Province. Additionally, it examines the interaction and integration characteristics of spatial form elements at three scales, courtyards, neighborhoods, and villages within rural settlement systems, from the bottom up. By combining appropriate resource utilization technologies and their basic resource metabolism data, this study set a resource metabolism target scenario of "complete organic waste elimination". This study then constructed a multi-scale resource metabolism flow model and spatial morphology model, proposing a spatial optimization strategy for rural settlements accordingly. The results and conclusions of this study strengthen the role of resource optimization in adjusting and guiding the spatial structure of rural settlements in China. They hold theoretical significance and practical value in promoting the integrated development of urban and rural areas, while broadening the concepts of rural construction design.

2. Overview of the Concept of Resource Metabolism in Rural Areas

Resource metabolism, originating from the biological term "metabolism", encompasses both "material metabolism" and "energy metabolism" [18]. The concept of resource metabolism in rural areas is predicated on the premise that a village functions as the boundary of the system. This includes the flow, accumulation, transformation, and other input and output processes related to the primary resources within the production and living systems of the area. Rural settlements can be conceptualized as human habitat spatial units with a fully integrated resource metabolism mechanism. In the early development of the field, resource metabolism was predominantly studied within urban contexts. American ecologist Abel Wolman initially introduced the concept of "urban metabolism", which integrates the flow processes of resource metabolism with the spatial elements of human settlements [1]. Currently, the theory of resource metabolism encompasses all scales, ranging from macro urban agglomerations to micro urban and rural settlements. Through interdisciplinary exploration, research on resource metabolism has developed a range of relatively mature analytical methods. Notably, the quantitative analysis method represented by material flow analysis (MFA), which quantifies mass fluxes of resource storage and flow conditions, has been employed to analyze the intrinsic characteristics and operational mechanisms of various resource flows within spatial systems and to elucidate the interactions between resource flows and the environment at specific locations. This approach elucidates the interactive relationship between resource flows and the environment at specific locations, facilitating the rational allocation and optimization of spatial resources. Such insights are significant for developing spatial construction strategies and methods for rural settlements aimed at the efficient utilization of resources.

The comprehensive recycling of resources has long embodied traditional ecological wisdom and remains a core concept in rural China. This principle aligns with the concepts and methods of urban resource metabolism applied in rural areas. Since the era of agrarian society in China, rural villagers have engaged in resource recycling and intensive land use within the framework of a self-sufficient small farm economic model. They have leveraged limited technology and space to develop various sustainable practices, including courtyard economy, household biogas production, and four-in-one facility agriculture. Numerous examples of eco-village construction have emerged, with Beijing's Illumining Village serving as a prominent representative. These villages have collectively amassed a wealth of invaluable experience [19]. From the perspective of resource elements and spatial composition, the configuration of rural settlements is more conducive to constructing an ideal resource metabolism and spatial modelling in rural areas. Firstly, the metabolic processes in rural areas are characterized by a comprehensive range of resource elements, a relatively straightforward spatial hierarchy, and a small scale. The agricultural space, forming the core link in the ago-ecological system, is capable of fulfilling multiple functions, including food production and the consumption and transformation of biomass resources. Furthermore, the metabolic processes in rural areas are based on three fundamental elements of resource metabolism: production, life, and ecology. Additionally, a complete cyclic path is in place, making it relatively straightforward to establish a resource production and utilization mode of "local resource recycling". Conversely, the spatial elements within the settlement are highly integrated, with production, living, and ecological spaces permeating and integrating with each other, and their boundaries are less defined. The spaces dedicated to agricultural production, water resource treatment, and human life are multifaceted and complex, making them more conducive to the integrated metabolic function of various resource elements in a complementary and symbiotic manner (Figure 1) [20].

Fruit and vegetable planting in front of village houses as a public productive landscape.

A lotus root planting pool with rainwater collection and purification functions is used as the village public space.

Figure 1. Integration of production and living space in rural residential areas.

3. Characteristics of the Interaction between Resource Metabolism in Rural Settlements and Spatial Elements

The layout of rural settlements is characterized by small and dispersed sites, complicating the direct application of traditional centralized spatial construction and infrastructure layout models to rural spaces. Findings from related research, exemplified by urban metabolism and agricultural resource management, have established a hierarchical recycling model of multiple resource mixing and a bottom–up distributed spatial operation method in terms of both technology and space [21]. These findings can effectively inform the study of resource metabolism characteristics in rural areas, the integration of spatial elements, and the rational allocation of spatial resources, in a manner closely aligned with the characteristics of agricultural resource utilization.

3.1. Spatial Layout Based on Resource Complementarity

The relative positioning of spatial elements in rural settlements necessitates comprehensive consideration of the economic and physical properties of resource elements. Enhanced spatial proximity of relevant functional facilities can ensure resource utilization efficiency and foster collaborative sharing [22]. Given the relatively loose and flat spatial characteristics of villages and the abundance of biomass resources, the distributed integration of resource treatment technology facilities with the spatial distribution of various living and production functions can reduce resource transmission and distribution losses. For example, biogas production facilities can be situated near courtyards and greenhouse planting facilities that require biogas. Small greenhouses and water resource collection systems can be integrated with courtyards, while solar water-heating devices can be installed on the sloping roofs of village buildings to provide domestic hot water or house heating. These arrangements enable the production, use, and recycling of resources in close proximity to the village, leveraging spatial proximity.

3.2. Bottom–Up Cascading Metabolic Flow of Resources

Resource metabolism in rural settlements operates at three spatial levels, the courtyard, the neighborhood, and the village, each of which can be structured to incorporate social, energy, food, and organic waste elements [23]. Interventions starting at the courtyard level are suitable for integrating resource utilization technologies, facilities, and spatial form elements. A bottom–up progressive approach allows resources that are not fully metabolized at lower levels to be transferred upward for recycling. This iterative process effectively reduces resource depletion, improves utilization efficiency, and ultimately controls and reduces the waste exported outside of a settlement system (Figure 2).

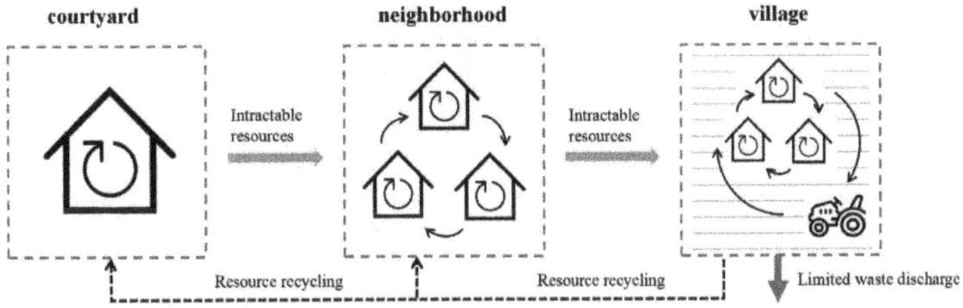

Figure 2. Bottom–up rural space optimization method.

3.3. Integration of Resource Metabolism Technology and Spatial Element Distribution

The distributed integration of appropriate resource utilization technologies and facilities with the spatial elements of rural settlements can facilitate resource recycling and preserve the rural landscape. This approach effectively reduces the input of manpower and material resources, and it better adapts to the decentralized and flat spatial layout of villages [24]. Some scholars have proposed strategies for optimizing the functional spatial layout of courtyards based on the "four-in-one" courtyard ecosystem operation model, which integrates biogas production and use [25]. A practical example is the Schoonschip design in Amsterdam, the Netherlands, which integrates a wide range of resource recycling technology facilities with small-scale grouped living spaces (Figure 3) [26]. Another example, the reconstruction project of Jintai Village in Sichuan, designed by Kenneth Lam of the University of Hong Kong, integrates various resource recycling technologies and facilities, such as livestock farming, rainwater collection, biogas production and use, and plant-based water purification systems, at the village unit and neighborhood levels. This project created a rural community space with efficient resource metabolism [27] (Figure 4).

Figure 3. Schoonschip ecological community in the Netherlands (available online at: https://schoonschipamsterdam.org/en/) accessed on 1 May 2024.

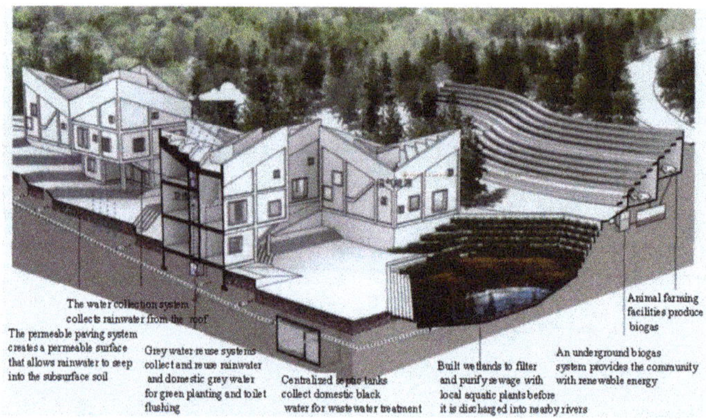

Figure 4. Resource metabolism system of spatial integration between Jintai Village and its neighbors in Sichuan Province (available online at: https://www.archdaily.cn/) accessed on 30 May 2024.

4. Research on the Methodology of Resource Metabolism Modelling

On the basis of elucidating the flow law of resource metabolism in rural settlements and its interaction with spatial patterns, resource metabolism in rural areas can be modelled based on the interdependence of resource elements and their quantitative flow dynamics, combined with the target state of resource metabolism in rural areas. The construction method of resource metabolism models has evolved from qualitative models based on the optimization of resource metabolism pathways to quantitative models grounded in the quantitative flow of resource metabolism accounting and spatial occupation.

Research on model construction based on quantitative resource flow metabolism accounting involves the quantitative analysis of the equilibrium-rationing relationship between the demand and supply of various types of resources at different process stages within a system, combined with the visualization of resource metabolism flow charts, to achieve quantitative model construction. Liu Changan constructed the resource metabolism model of an "agricultural community", based on rural production and nutrient metabolism, optimizing and improving the resource cycle system of the associated production and living space at the scale of an urban community [28]. Some scholars have employed the system dynamics method to establish a resource metabolism model for eco-villages under the zero-waste paradigm by measuring the quantitative correspondence between resource consumption and spatial occupation, based on the "3R" principles of resource reduction, reuse, and recycling (Figure 5) [29].

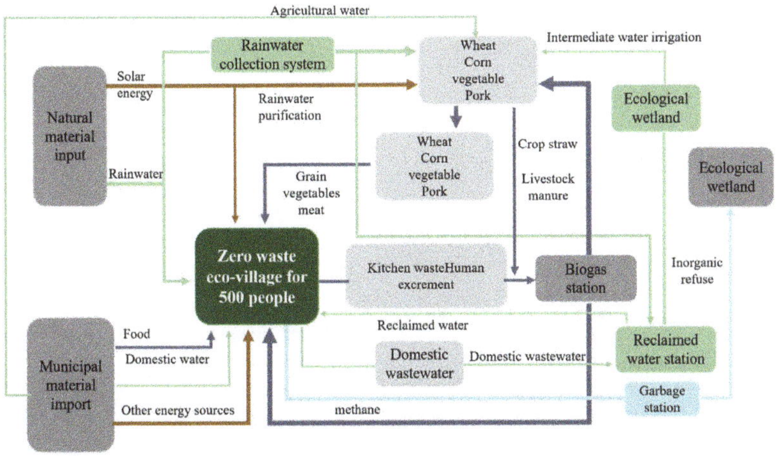

Figure 5. Resource metabolism modelling for "zero-waste" eco-villages.

Research on the construction of spatial occupancy models based on the quantitative metabolic flow of resources involves measuring and establishing a quantitative model of a functional spatial land use scale to achieve balanced resource metabolism rationing, grounded in the analysis of the resource demand, transformation, and output of relevant spatial elements. For example, Edward Yiu of the Chinese University of Hong Kong, through quantitative calculations of the cyclic metabolism of energy, water, food, and organic waste, demonstrated that 40 hectares of land were required to satisfy the food demand of 10,000 people in a zero-waste community. He established a novel planning model aimed at creating a zero-food and zero-organic waste community (ZFW community) [30]. In the Synergy between Regional Planning and Exergy (SREX) Project, Wageningen University in the Netherlands introduced the concept of "Urban Tissue", which is based on the utilization of energy and water resources. The "Urban Tissue" concept uses a 1-hectare land area (100 m × 100 m) as a proposed spatial unit. Its goal is to achieve a closed cycle of resources, using per capita resource consumption and waste emissions as basic data. This concept determines various types of functional space and land area indicators to construct a metabolic spatial model of the urban functional unit [31] (Figure 6). At the scale of rural settlements, the Netherlands' "Regen Village" project, based on the concept of resource metabolism, integrated community size with the per capita consumption of major resources. This project established a small ecological community resource metabolism model covering an area of 15,500 m^2 to meet the basic living needs of 60 residents, by combining appropriate functional space layouts and technical facilities for resource use. Japanese scholars, such as Takeuchi, regard rural space as an area capable of self-sufficiency. By combining the characteristics of rural resources with the flow of urban and rural factors and considering the distance from urban areas, they constructed three types of ecological rural spatial models: suburban villages, peri-urban villages, and natural villages, each with a self-cycling resource metabolism mechanism [32] (Figure 7).

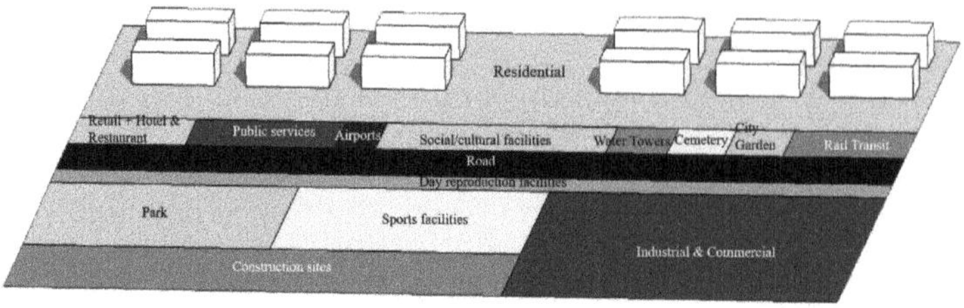

Figure 6. Conversion of quantitative metabolic values for resource metabolism balancing categories in the "Urban Functional Unit" into land area indicators.

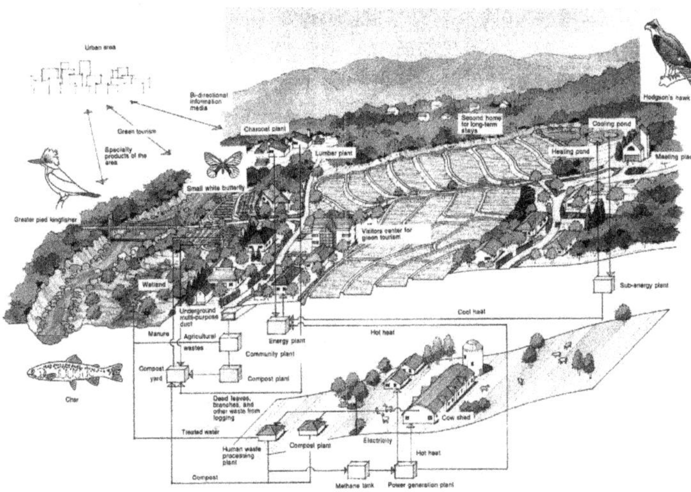

Figure 7. A spatial model of a Japanese eco-village based on the concept of in situ recycling of resources (source from: reference [16]).

5. Modelling of Resource Metabolism in Rural Settlements in the Shandong Plain Area

5.1. Characteristics of the Habitat Environment and Current Status of Resource Utilization in the Shandong Plain Region

The plain terrain of Shandong Province comprises approximately 55% of the province's total land area [33]. Additionally, the number of villages, the size of the agricultural population, and the economic volume in these plain areas are proportionally higher [34]. These settlements, representing the quintessential rural habitat of Shandong, are characterized by relatively complete infrastructure and public service facilities. They are anticipated to survive and develop over the long term and thus are a central focus of both current and future rural development efforts in Shandong. Although the villages in Shandong's plain areas have a relatively plain character, they benefit from convenient transportation and location. This proximity fosters closer interactions with urban areas in terms of economy, population, space, and resources. Consequently, these villages are more susceptible to urban influences, exacerbating the conflict between human settlement development and resource utilization [35]. Combined with a large number of case field studies in plain villages, a large amount of basic information looking into ways of utilizing the four types of resources associated with the living and production systems, namely water, organic waste, energy, and agriculture, has been collected (Figure 8), which mainly contained the following characteristics.

Figure 8. Types of resource metabolism in research cases.

With regard to water resource utilization, the penetration of piped water facilities in rural Shandong Province has achieved 97% coverage [36], Nonetheless, numerous farmers continue to rely on personal or communal wells due to the costs associated with piped water. Furthermore, sewage collection and treatment are predominantly managed by small-scale centralized facilities. In some villages, ecological sewage treatment is integrated with landscaping and agricultural practices; however, overall water use remains relatively unsophisticated, with insufficient recycling and tiered utilization. Regarding organic waste treatment, plain villages are abundant in biomass resources. Most villages have implemented on-site resource utilization strategies, converting agricultural organic waste into straw biogas or biomass fuels, thereby enhancing the connection between production and living systems. However, the sorting and recycling rates of domestic organic waste remain inadequate. Concerning energy use (heat and electricity), with the exception of a few peri-urban villages employing centralized heating systems, most villages depend primarily on ambient heat or electricity for winter heating. Additionally, some villages utilize centralized biogas stations or small household digesters for biogas production for cooking purposes. Distributed solar photovoltaic and solar thermal systems, integrating with the roofs of village houses and unused land, are increasingly prevalent. Regarding agricultural production, traditional large-field cultivation, greenhouse planting, and large-scale farming are predominant methods in plain villages. Continuous greenhouses have emerged as prominent features in the rural landscape on the periphery of some villages. These practices utilize both courtyard areas and previously underutilized vacant land for small-scale "production and residence integration" farming methods, thereby contributing to a distinctive productive landscape.

5.2. Resource Metabolism Spatial Modelling Path

To investigate the mechanism of resource metabolism in rural settlements, it was imperative to examine both resource and spatial elements comprehensively. Resource metabolism in rural areas begins with fundamental material elements that sustain the normal function of the rural environment. It involves analyzing the mechanisms of nutrient, water, and energy cycles within the community, with particular emphasis on the nutrient cycle, which focuses primarily on food production and the cyclic metabolism of organic waste [37]. The resource elements addressed in this study, excluding inorganic waste (e.g., solid waste and old building material) that are challenging to recycle effectively within the spatial and technological constraints of villages, encompass four primary categories: water, energy (both electrical and thermal), food (agricultural production), and organic waste (nutrients). Regarding spatial elements, the typical rural spatial framework comprises four main categories: built-up elements (roads, courtyards, public buildings), natural environmental components (vegetation, water bodies), agricultural production elements (arable land, greenhouses), and infrastructural components (electric power transmission and distribution facilities, water and sewage systems). These elements are interwoven and integrated at various scales. The courtyard comprises the residence and its ancillary functional rooms, hardscaped areas, planted vegetable gardens, small greenhouses or livestock spaces, integrated solar photovoltaic and solar thermal facilities on roofs, walls, or underground, rainwater catchment areas, and organic waste collection ponds (septic tanks). The neighborhood extends beyond basic courtyard elements to include the road in front of residences, public spaces, and areas where resource metabolism can be integrated, such as agricultural greenhouses, open vegetable gardens, small ecological wetlands, breeding spaces, and underground biogas facilities. The village encompasses additional functional elements outside the courtyard, including rural public service buildings, large public open spaces, medium- to large-scale biogas and sewage treatment stations, agricultural land, extensive greenhouses, and various natural water bodies. Adhering to the principle of achieving balanced resource metabolism across varying spatial scales, this study approached the task from three hierarchical levels—the courtyard, neighborhood, and village. Following the framework of "resource metabolism data and spatial indicator calculation—resource metabolism scenario setting—balanced rationing model analysis—resource metabolism and spatial model construction", this study developed a comprehensive resource metabolism model and spatial morphology model incorporating agricultural production factors, ecological and technological facilities, and human living systems (Figure 9).

(1) Data Accounting

Data accounting encompasses both resource and spatial data related to resource metabolism in rural areas. Concerning resource metabolism data, basic information on the rural per capita or household consumption within the agricultural system, ecological facility system, and human living system was obtained from sources such as the China Statistical Yearbook, the Shandong Statistical Yearbook, and the China Household Energy Consumption Research Report. These data, covering the metabolic flow of four primary resource elements—water, food, energy, and organic waste—formed the foundational basis for constructing the resource metabolism model. By integrating spatial occupancy data for open land and greenhouses used in the production of major crops and livestock under the typical environmental and climatic conditions of the Shandong Plain, along with spatial data for land or buildings equipped with appropriate resource processing technologies, the basic spatial data indices for resource metabolism in rural areas were established.

(2) Situation Setting

Compared to the resource metabolism in urban areas, which primarily involves a unidirectional input of resources, plain-type rural settlements generally function as resource output systems due to their agricultural production roles. In these settlements, the volume of agricultural output typically exceeds the consumption levels within the villages (considering only the essential food requirements of residents and livestock). Therefore, addressing the villages' self-sustaining development needs and habitat remediation goals—focusing on the utilization of organic waste resources—and considering the village development model and spatial resource needs required the establishment of different resource metabolism scenarios. This included evaluating various types of resource metabolic flows, balanced ration data, quantitative resource flow, the degree of functional integration at different scales, resource self-sufficiency rates, and spatial element diversity.

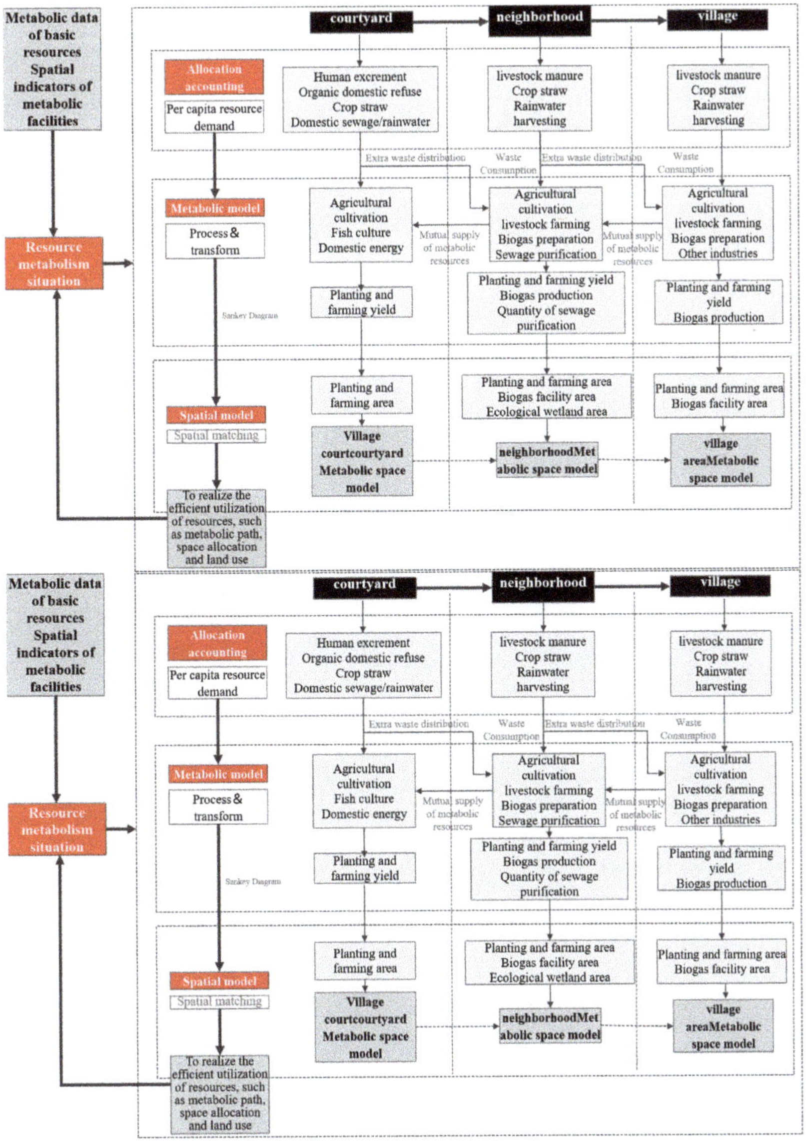

Figure 9. Resource metabolism in rural areas and spatial model construction ideas.

(3) Modelling

Quantitative Metabolic Analysis. By integrating the interaction characteristics of appropriate resource metabolism technologies and rural spatial elements, and utilizing resource metabolism quotas derived from human resource consumption data, this study performed a bottom–up analysis of resource metabolism objectives at three spatial scales—the courtyard, neighborhood, and village. This analysis considered the capacity of ecological technology systems (resource metabolism facilities) applicable to these scales. This study assessed the metabolic flow of major resources and their balanced quota relationships within and between different scales, employing Sankey diagrams for visualization. This balanced quota analysis helped determine resource flow statuses and established a model of resource metabolism in rural areas.

Spatial Modelling. Constructing a spatial model for rural resource metabolism required integrating the abstract resource metabolism model with the tangible spatial morphology of the village. This involved translating quantitative resource flow data at various scales into categorized spatial morphology indices which included relevant technological facilities. For example, this included the area occupied by photovoltaic installations on village roofs, road space used for domestic wastewater transmission, and courtyard or neighborhood open spaces allocated for small greenhouses. By converting quantitative resource indicators into spatial form indicators, this study could develop a spatial model that achieved balanced resource metabolism within the village boundaries (Figure 10).

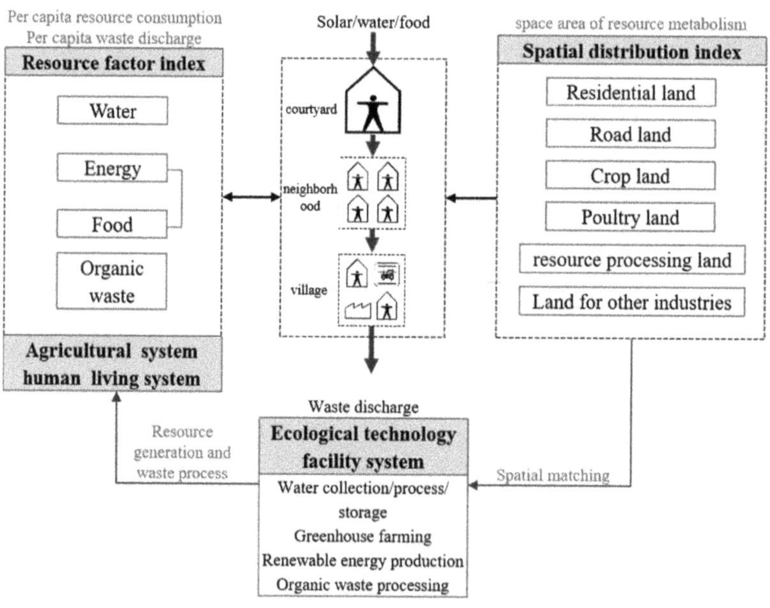

Figure 10. "Resource-space" rationing of rural human space.

5.3. Setting of Relevant Technical and Spatial Indicators for Resource Metabolism Systems in Rural Areas

The resource metabolism in the rural areas was quantitatively assessed with a focus on the needs of individuals, addressing their fundamental production, living, and ecological requirements. The analysis categorized resource metabolism into three subsystems: the human living system, the agricultural production system, and the ecological and technological system. This assessment was informed by data from statistical yearbooks of China and Shandong Province, along with the relevant literature [38,39]. Basic consumption data for the four primary resources—water, energy (both thermal and electrical), food (planting and breeding), and organic waste—were compiled separately. Additionally, data on the production and conversion of resources through agricultural and ecological technologies (such as biogas production, solar photovoltaic systems, and ecological sewage treatment) were integrated to achieve balanced resource rationing. This approach provided the foundation for constructing robust resource metabolism models.

5.3.1. Basic Data on the Resource Elements of the Three Living Systems

Given that this study aimed to achieve metabolic spatial modelling through quantitative resource metabolism analysis, the focus of data screening was not on pursuing exact precision but on establishing the reasonable availability of data. According to the 2021 Shandong Statistical Yearbook, the average number of permanent residents in Shandong's rural areas was 3.1 per household. Consequently, resource consumption for courtyard households was calculated based on this average, and resource quantities such as those related to livestock breeding were adjusted in accordance with the general population of villages and the spatial scale of courtyard areas (Tables 1–3).

Table 1. Resource element ratios for human living systems.

Resource Type			**Use**	**Annual Resource Consumption per Capita**	**Average Annual Resource Consumption per Household (Calculated for 3.1 Persons)**	**Marginal Notes**
Resource input	Energy	Electric	Living electricity	714 kwh/yr	2213 kwh/yr	Included water for drinking, washing, cooking, etc.
		Heat	Cooking	106 m^3/yr (biogas)	317 m^3/yr (biogas)	
			Heating	61 kg/yr	184 kg/yr	
			Total amount of energy	135.06 kg/person/yr	405 kg/person/yr	
	Water		Domestic water use	24.6 m^3/yr	76 m^3/yr	Included water for drinking, washing, cooking, etc.
			Natural precipitation		114 m^3/household/yr	The average precipitation in Shandong was 710 mm/year, and the area of precipitation collection per household was calculated according to the average household's hard interface area of 160 square meters
	Food needs		Cereal consumption	168.4 kg/yr	522 kg/yr	Contained grains, potatoes, beans
			Meat consumption	27.4 kg/yr	85 kg/yr	Reference to the China Statistical Yearbook (2020) and China Food and Nutrition Development Program (2014–2020)
			Fish consumption	11.2 kg/yr	34.72 kg/yr	
			Vegetable consumption	140 kg/yr	434 kg/yr	
Resource output	Organic waste		Human excreta	164.25 kg/yr	509 kg/yr	
			Domestic sewage	21.4 m^3/yr	66 m^3/yr	
			Organic waste	182.5 kg/yr	566 kg/yr	

Table 2. Resource factor rations for agricultural production systems.

Resource Type				Average Annual Resource Production	Average Household Resources	Volume of Resources for the Cluster	Volume of Resources in the Village Area	Notes
Resource input	Nutrients	Greenhouse		Biogas: 3.5 kg/m^2/yr Residue: 3 kg/m^2/yr	–	–	–	1. Priority was given to the use of rainwater harvesting for watering agricultural plantations, and any shortfalls were mainly imported from outside the system using municipal pipeline networks or river and lake water bodies 2. The use of greenhouse intensive cultivation on the roofs of village houses and public sites of neighboring groups could produce 12.69 kg/m^2/year (8452 kg/mu/yr) of vegetables per square meter of the rooftop greenhouse 3. Feeding cycles: 365 days for cattle and sheep, 180 days for pigs, 50 days for broilers 4. Needed to be measured in relation to specific acreage
		Open-air vegetable plots						
		Cropland		Biogas: 5000 kg/mu/yr Biogas: 3750 kg/mu/year				
	Food	Pigs		Digestate: 1460 kg/pc				
		Poultry		36.5 kg/pc				
Resource output	Food and vegetables	Greenhouse vegetables		8452 kg/mu	–	–	–	
		Open-air garden		3330 kg/mu				
		Cropland grains		416 kg/mu				
	Meats	Hog		70 kg/head	–	11 heads	–	
		Meat and poultry		3.5 kg/pc		13 pc		
		Cow		250 kg/pc		–	17 pc	
		Goats		20 kg/pc			186 pc	
	Caviar	Pond aquaculture		11.2 kg/yr	34.7 kg/household/yr	–	–	
	Organic waste	Pig manure		1930 kg	–	21,230 kg	–	
		Poultry manure		36.5 kg		475 kg		
		Cattle manure		9855 kg		–	167,535 kg	
		Sheep manure		865 kg			160,890 kg	
		Quantity of straw	Greenhouse vegetables	740 kg/mu (1.54 kg/m^2)		–	–	
			Open-air garden	633 kg/mu (0.95 kg/m^2)				
			Grains	376 kg/mi				

Table 3. Eco-technology system resource element rations.

Resource Type	Use		Resource Handling Capacity	Appropriate Scale			Notes
				Courtyard	Neighborhood	Village	
Electricity	Solar photovoltaic installations		262.8 kwh/m^2/yr	•	•	•	1. Based on an average of 6 h of effective light per day 2. Rooftop greenhouses integrated with village buildings could transport some of the heat to the building interior 3. Processing 1 m^3 (tons) of domestic wastewater required an approximate footprint of 1.7–2 m^2 4. Treatment of 10,000 m^3 of sewage per day, with a treatment facility covering an area of approximately 17,900 square meters 5. According to the calculation of a 6 m^3 small-scale biogas digester, the annual gas production was about 288.72 m^3, and then the annual biogas production of the 1 m^3 biogas digester was 48.12 m^3 6. The biogas production of animal manure was 0.06 m^3/kg, the biogas production of domestic organic waste was 0.1 m^3/kg, and the biogas production of straw was 0.28 m^3/kg 7. A total of 1 m^3 biogas could produce 3.42 kg of digestate and 2.28 kg of digestate 8. Average annual consumption of organic waste per unit area of cultivation space
Heat energy	Solar thermal facilities		304 kwh/m^2/yr	•	•		
	Rooftop greenhouse		255 kwh/m^2/yr	•			
Sewage treatment	Small-scale artificial ecological wetland sewage treatment		2 m^2/m^3	○	•		
	Artificial sewage treatment plant		17,900 m^2 = 10,000 m^3 wastewater treatment capacity			•	
Organic waste treatment	Landfill gas	Household biogas facilities	48.12 m^3/yr		•		
		Medium and large biogas plants	418.75 m^3/m^3/yr			•	
	Plantation	Small-scale greenhouse cultivation	Organic waste 1.53 kg/m^2/yrWastewater 2.92 m^3/m^2/yr	•	•		
		Open-air garden planting		•	•		

Clarification: • indicates the optimal scale for a resource metabolism technology facility, ○ indicates a more appropriate scale.

5.3.2. Spatial Hierarchy of Rural Settlements and Indicator Setting

The balance quota accounting of the resource metabolism system, which involved tracking the flow of various resource elements at different scales within the village, had to be integrated with rural community planning and relevant technical specifications. This integration included setting spatial indicators for the three scales, the courtyard, neighborhood, and village, and defining their resource metabolism spatial layouts. This process provided a spatial operational basis for the subsequent transition from quantitative resource metabolism indicators to spatial morphology indicators of resource metabolism.

(1) Courtyard

Spatial indicator setting. According to the Technical Guidelines for the Construction of New Rural Communities in Shandong Province (for trial implementation), the land control indices for village settlements (Table 4) indicated that the average residential land area in the plain regions was 200 m^2 per household. Combining this with the 2021 statistical data showing an average per capita residential floor area of 43.41 m^2 for rural residents in Shandong Province, the average household floor area was approximately 135 m^2. This figure was used as the minimum spatial index for the village building model [40]. Regarding basic spatial form control, current regulations stipulate that rural settlements should feature 2–3-story low-rise townhouses and multi-story residential units not exceeding six stories. Considering the traditional rural landscape, which primarily consists of low-rise buildings, the proposed model for courtyard spaces utilizes low-rise townhouses as the fundamental residential units. The architectural design combines flat and sloping roofs, with flat roofs for the south-facing main houses and sloping roofs for the main rooms facing south. Auxiliary rooms on the east and west have flat roofs. This design reflects the traditional architectural style of Shandong Province's plains while accommodating ecological and technological facilities, such as solar photovoltaic panels and small greenhouses. The integration of these elements with courtyard walls creates an open space suitable for family activities, outdoor vegetable gardens, family farming, and resource metabolism facilities like rainwater collection.

Table 4. Spatial indicators for courtyards (basic spatial units).

Type of Village	Households per Capita Homestead	Average Household Floor Area	Story	Plot Ratio
Plain-type villages	≤200 m^2	135 square meters	2–3 floors	≥0.3

Note: This indicator was calculated on the basis of the land control indicators for the construction of village settlements in the Technical Guidelines for the Construction of New Rural Communities in Shandong Province (for trial implementation), combined with the data from the Statistical Bulletin on National Economic and Social Development of Shandong Province in 2022.

Space Usage Settings. Based on the aforementioned spatial control indices and the morphological characteristics of low-rise courtyards and physical buildings in the Shandong Plain, the spatial allocation and area index range for resource metabolism in rural courtyards were established. To meet basic access requirements, a small rooftop greenhouse, approximately 24 m^2 in area, was installed on the roofs of the east and west rooms on the first floor. An open vegetable garden, not exceeding 30 m^2, and a pond (fish pond), not exceeding 8 m^2, were positioned in the adjacent open space. Additionally, solar photovoltaic or solar thermal panels were installed on the south side of the sloping roof of the main house, sized according to the energy needs for living and production. These spatial and quantitative indices for various resource metabolism facilities are outlined, with specific ranges to be refined through subsequent metabolism accounting under different resource metabolism scenarios. The greenhouse planting and open vegetable gardens were designed to be adaptable and replaceable (Figures 11 and 12).

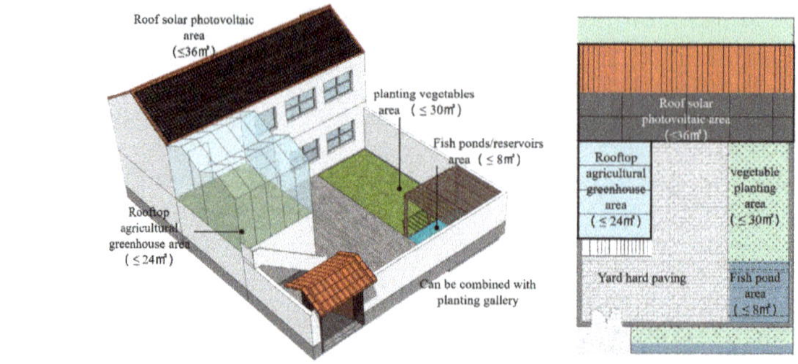

Figure 11. Basic model of the courtyard and setting of the spatial use model.

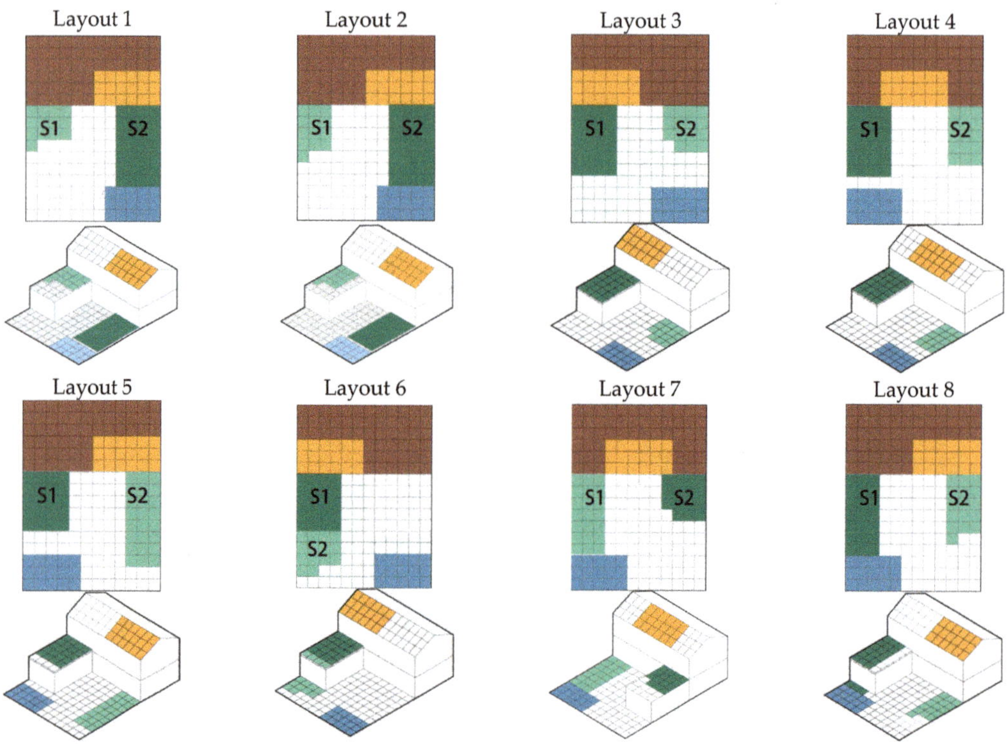

Description: 200 m^2 (16 m × 12.5 m) of residential area, 135 m^2 of building area of village house, S1 is greenhouse planting combined with the roof of the village house, S2 is the ground vegetable garden in the courtyard, yellow is solar photovoltaic and solar thermal panels combined with the south-facing sloped roof, and blue is the water-collecting fish pond in the courtyard.

Figure 12. Schematic layout of the courtyard to realize the balance of resource metabolism.

(2) Neighborhood

Spatial Indicator Setting. Neighborhoods in rural settlements are medium-scale spatial aggregates comprising multiple courtyards, public spaces, and community facilities. These clusters serve as crucial links between individual courtyards and the broader spatial-scale resource metabolism. The number of courtyards within a neighborhood must be evaluated from both technical and social perspectives. Technically, the use of "household biogas" systems for balancing resource production and consumption provides a key basis for determining the appropriate scale of neighborhoods. By integrating technology and spatial considerations, and minimizing the distance for biomass resource transfer, the total amount of organic waste input and biogas output could be calculated to estimate the number of courtyards within a cluster. Practical examples and feedback from research supported this approach: for instance, 32 households in Jiangfang Xincun, Yongyang townships shared a biogas digester and an 8 m^3 storage tank; seven households in Longpi Village shared a 10 m^3 digester; and eight households in the EVA-Lanxmeer settlement in the Netherlands had a biogas production unit for every eight households [41]. Additionally, considering the social dynamics of "acquaintance societies", where neighborly interaction and a sense of community are significant, research indicates that residential clusters of 8–12 households foster a positive social atmosphere [42,43]. Thus, balancing both technical and social factors, the recommended number of courtyards in a neighborhood was set at 10 (Figure 13).

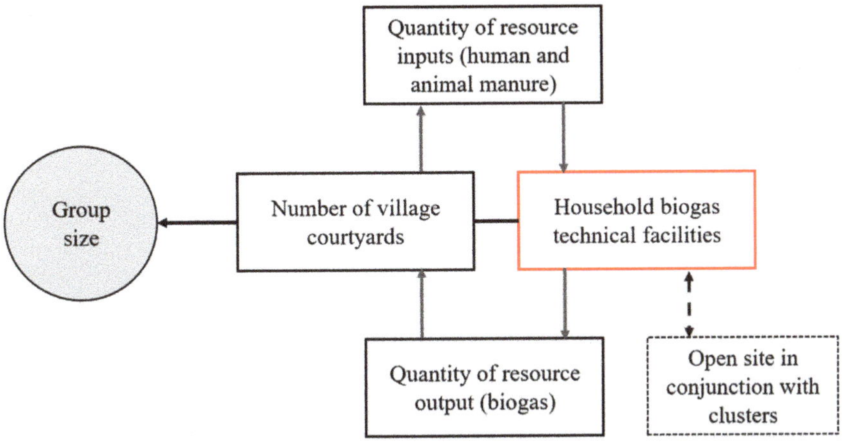

Figure 13. Flow chart outlining the determination of neighborhood sizes.

Space Use Mode Setting. Taking as an example the proposed enclosed neighborhood composed of 10 courtyards, the spatial components included two main categories: courtyards and neighborhood open spaces. Besides the road interface, which was dedicated to precipitation collection, the remaining areas between the courtyard walls and the road, as well as the open spaces (primarily unused vacant or residential land), could be utilized for various purposes. These included vegetable gardens, greenhouses, livestock breeding (such as pigs and poultry), and small-scale artificial wetlands integrated with hydroponics and vegetable cultivation (encompassing water transfer and distribution ditches, anaerobic ponds, regulating ponds, ecological oxidation ponds, and other components). The spatial indices for resource metabolism, excluding the road (which covered approximately 10–15% of the total area) and the roadside productive greenery (around 235 m^2), should be precisely determined based on the specific metabolic targets for each resource through rationing calculations (Figure 14). Additionally, for the 10 courtyards and public open spaces, different combinations of organic waste metabolism technologies should be integrated. Depending on the current situation, these spaces can be fully occupied and utilized for various scales and types of unused open areas, leading to different layout patterns such as enclosing groups, rows, and columns, to accommodate or develop diverse village morphologies (Figure 15).

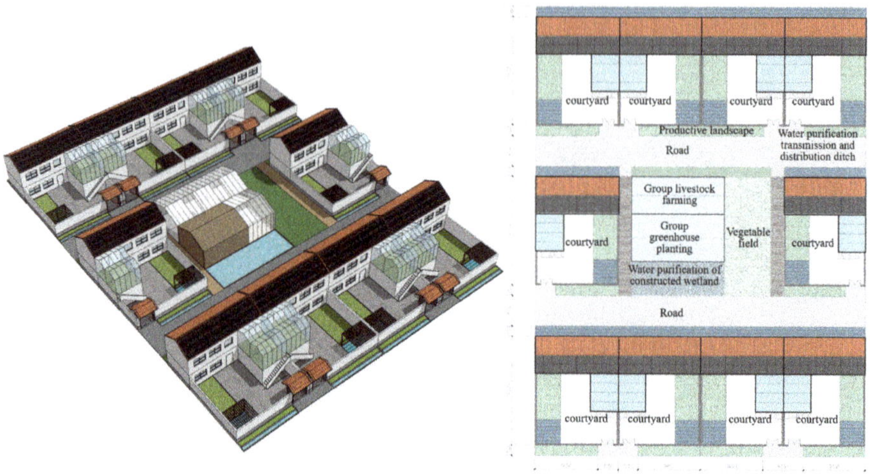

Figure 14. Basic layout setting for a neighborhood space.

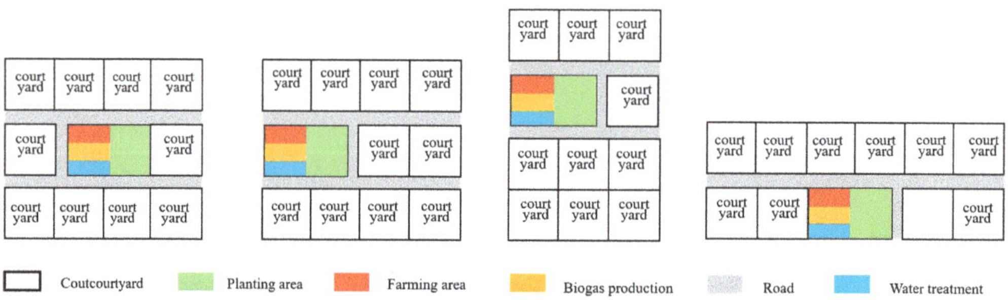

Figure 15. Schematic illustration of the metabolic spatial layout approach for neighborhoods.

(3) Whole Village

Area Allocation of Spatial Elements. To facilitate the measurement of overall resource metabolism within a village, it was essential to determine the village size and land use indicators. The village scale was based on population or household numbers. Using data from Shandong Province, where the average village population was 530 and the average household size was 3.1 in 2020, it was estimated that each village had approximately 170 courtyards. With a basic residential land area of 200 m^2 per household, the total residential land area amounted to about 34,000 m^2. Given the per capita arable land in Shandong Province was approximately 1.21 mu (3.75 mu per household), the total agricultural land within the village was estimated to be around 638 mu (424,575 m^2). Dividing the village into groups of 10 households each resulted in 17 neighborhood spaces. Public buildings and green spaces should accommodate centralized resource metabolism functions, while other spaces should be allocated based on quantitative resource metabolism accounting results.

5.4. Resource Metabolism Modelling and Spatial Modelling in Rural Settlements—Based on the "Organic Waste Self-Consumption" Situation

The study of the resource metabolism in the rural settlements within the Shandong Plain was based on the fundamental data of resource metabolism and considered the current production, living conditions, and resource utilization in these areas. This study established a resource metabolism scenario focused on "organic waste self-consumption", integrating this with the concepts of resource metabolism and the "Production-Living-Ecology" framework. This scenario allowed for the analysis of resource metabolism flows and the calculation of balanced quantities within the village. By incorporating these concepts, this study analyzed resource metabolism flows and balanced rationing. Subsequently, this study constructed a resource metabolism model and a spatial pattern model from the ground up to optimize and enhance the resource metabolism system in rural settlements, improving its integration with spatial elements.

The framework of "organic waste self-consumption" in resource metabolism was established, ensuring that while meeting the fundamental food needs of the village population and livestock, organic waste such as sewage, feces, and straw generated from human, livestock, and agricultural activities was systematically utilized. This approach addresses the primary issues concerning the ecological improvement in rural environments today. This approach tackles critical issues in rural ecological enhancement, encompassing sewage management and local organic waste resource utilization. It is crucial to note that the aim of resource metabolism equilibrium is to maximize the metabolic cycle and self-sufficiency of resource elements within a system. However, this does not imply a completely closed-loop metabolic system without external inputs. Firstly, the rural resource metabolic system requires the continuous external input of solar energy and water resources, including municipal water sources and natural precipitation, as solar energy is essential for all metabolic activities. Secondly, energy self-sufficiency does not imply 100% self-production and consumption but aims for a quantitative balance between energy production and utilization within the system. Lastly, the fertilizers and elements needed for crop production cannot be entirely self-sufficient and must be supplemented externally.

5.4.1. Accounting for the Resource Metabolism Base Allocation

Basic Resource Requirements: Using the smallest spatial unit—a courtyard with an average of 3.1 persons per household—as the base parameter for resource consumption, Table 5 illustrates the annual consumption of major food and water resources and the production of organic waste at the courtyard (3.1 persons), neighborhood (31 persons), and village (530 persons) scales. The resource metabolism capacities and spatial elements at these three scales, along with appropriate resource metabolism technologies, varied. Consequently, the organic waste produced was not directly proportional across scales. For example, in courtyards, organic waste primarily consisted of human feces, domestic organic waste, domestic sewage, and planting organic waste. Small greenhouses, open-air vegetable gardens, and fish farming ponds could consume some of this waste. At the neighborhood scale, additional fecal organic waste from pig and poultry farming was managed, utilizing household biogas tanks and small-scale artificial ecological wetlands to process animal organic waste, domestic sewage, and rainwater. At the village scale, organic waste increased due to the large-scale breeding of cows and sheep, necessitating medium- and large-sized biogas stations for centralized digestion and treatment.

Table 5. Biomass resource base to be consumed (annual average).

Scale Levels	Human and Animal Waste			Domestic Organic Waste	Water		Crop Straw	Notes
	Human	Pig + Poultry	Cow + Sheep		Domestic Sewage	Rainwater		
Courtyard	509 kg	–	–	566 kg	66 m^3	114 m^3	Calculated on the basis of crops that consumed organic waste	Courtyard:neighborhood:village = 3.1:31:530
Neighborhood	5090 kg	21,705 kg	–	5660 kg	660 m^3	345 m^3		
Village	86,530 kg	–	134,228 kg	96,220 kg	–	–		

5.4.2. Resource Metabolism Modelling

Effective organic waste metabolism and consumption require appropriate transformation and utilization of various biomass resources, following the spatial hierarchy of "courtyard-neighborhood-village" from the bottom up. According to the spatial hierarchy of "courtyard-neighborhood-village" and the principle of "integrating production and life, metabolizing and consuming locally", it was essential to measure the resources required for metabolizing and transforming organic waste at the courtyard scale, along with the spatial indices of various productive spaces and appropriate resource metabolism technologies.

(1) Courtyard

Courtyard space primarily functions for water storage and managing the output of human excreta and organic waste, which is then transported and distributed to neighborhoods. A total of 114 m^3 of collected rainwater directly supplemented fish pond water, followed by the irrigation for fruits and vegetables in courtyards, with any shortfall made up by municipal pipeline water. Based on a 10-day water exchange frequency for fish farming, an ideal pond area of 6.3 m^2 (with a depth of 0.5 m) could meet the annual demand of 34.7 kg of fish per household. Considering gradient utilization and odor issues, 509 kg of human excreta and 566 kg of domestic organic waste were transferred to neighborhood-scale household biogas facilities for oxidative fermentation, generating 87.14 m^3 of biogas for village cooking energy (fecal biogas: 0.06 m^3/kg × 509 kg = 30.54 m^3; domestic organic waste biogas: 0.1 m^3/kg × 566 kg = 56.6 m^3). This process also produced 198.7 kg of methane and 298 kg of digestate, prioritized for fish farming and crop cultivation. Integrating the courtyard space, fish farming required 35 kg/year of methane liquid. The courtyard could support 24 m^2 of rooftop greenhouse and 30 m^2 of an open-air vegetable garden, needing 3.5 kg/m^2/year of methane liquid and 3 kg/m^2/year of methane residue. Methane liquid was fully utilized for fish farming and garden planting, with some residues used for planting. The remaining 276 kg of residues were exported to the neighboring group space. A 24 m^2 rooftop greenhouse could produce 305 kg of vegetables annually (12.7 kg/m^2 × 24 m^2). The remaining digestate was used in the neighborhood space. The 30 m^2 open-air vegetable garden produced 150 kg annually (5 kg/m^2 × 30 m^2), resulting in a total of 455 kg of vegetables and 65.46 kg of straw per year (1.54 kg/m^2 × 24 m^2 + 0.95 kg/m^2 × 30 m^2 = 65.46 kg). These vegetables met the annual demand of 434 kg per household, achieving a vegetable self-sufficiency rate of 105%. Straw was delivered to the group level to feed livestock (Table 6, Figure 16).

Table 6. Resource metabolism status and metabolic modelling at the courtyard level (3.1 persons/household).

Metabolic State	Resource Type	Indicator (yr)	Clarification
Resource input	Rainwater harvesting	114 m^3	Municipal domestic water supply was not included in the calculation process
	Biogas	264.6 m^3	Biogas was the total amount of biogas from both the neighborhood and village scales, and digestate was the total amount of resources to satisfy the cultivation of fruits and vegetables in gardens
	Biogas slurry	198.7 kg	
	Biogas residue	7.22 kg	
Internal metabolism (biology)	Vegetable growing	455 kg	Yard gardening for self-sufficiency
	Aquaculture	34.7 kg	
Resource output	Human waste	509 kg	Distribution to neighborhood spaces for metabolic treatment use
	Domestic sewage	66 m^3	
	Domestic organic waste	566 kg	
	stalk	65.46 kg	

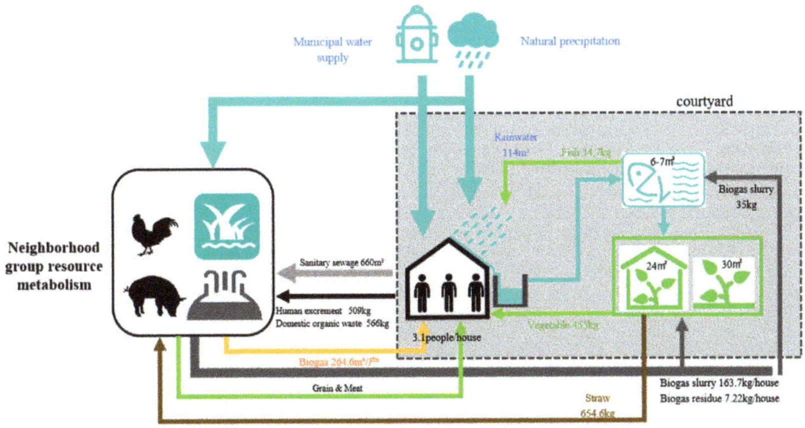

Figure 16. Resource metabolism modelling of courtyards.

(2) Neighborhood

The neighborhood primarily managed livestock manure and domestic sewage in the communal space. With courtyards already meeting vegetable demands, group plantings focused on cereals. To meet the annual meat demand of 10 villagers, 11 pigs and 13 meat birds were raised in the group's communal area, producing sufficient meat and generating 21,705 kg of manure. Approximately 500 m^2 of green space was used for cereal cultivation, yielding 282 kg of cereals annually (28.2 kg per household) and 312 kg of straw. The total straw production from 10 households' vegetable gardens was 967 kg (655 kg from gardens + 312 kg from the cluster), used for livestock breeding. The cluster's household biogas facilities processed 26,795 kg of manure (21,705 kg from livestock + 5090 kg from villagers) and 5660 kg of domestic waste, producing 2173 m^3 of biogas. The biogas residue from 21,705 kg of livestock manure, combined with 2760 kg of residue from 10 village gardens, totaled 7214 kg of biogas residue. The 500 m^2 cereal planting area within the cluster consumed 2969 kg of biogas and 586 kg of digestate. The remaining 6628 kg of digestate was transferred to the village area for centralized agricultural use (Table 7, Figure 17). The resource metabolism flow within the courtyard system boundary is illustrated through a Sankey diagram, reflecting the integration of resource metabolism processes with the spatial morphology of the village, facilitating subsequent spatial metabolism modelling.

Table 7. Resource metabolism status and metabolism models at the neighborhood level (10 households/groups, 31 persons).

Metabolic State	Resource Type	Indicator (yr)	Clarification
Resource input	Rainwater harvesting	412 m^3	Municipal water supply was not included in the calculation process
	Human waste	5090 kg	From the courtyard
	Domestic sewage	660 m^3	
	Domestic organic waste	5660 kg	
	Fruit and vegetable stalks	655 kg	
	Livestock breeding feed	15,568 kg	The amount of external straw feed still needed to be supplied

Table 7. *Cont.*

Metabolic State	Resource Type	Indicator (yr)	Clarification
Internal metabolism (biology)	Grain stalk	312 kg	Group cereal farming
	Animal manure	21,705 kg	–
	Digestate	2969 kg	
	Digestate	586 kg	
Resource export	Cereal cultivation	282 kg	Distribution to courtyards
	Poultry and meat	706 kg	
	Landfill gas	2173 m^3	
	Digestate	1987 kg	
	Digestate	72.2 kg	
		6628 kg	Distribution to villages

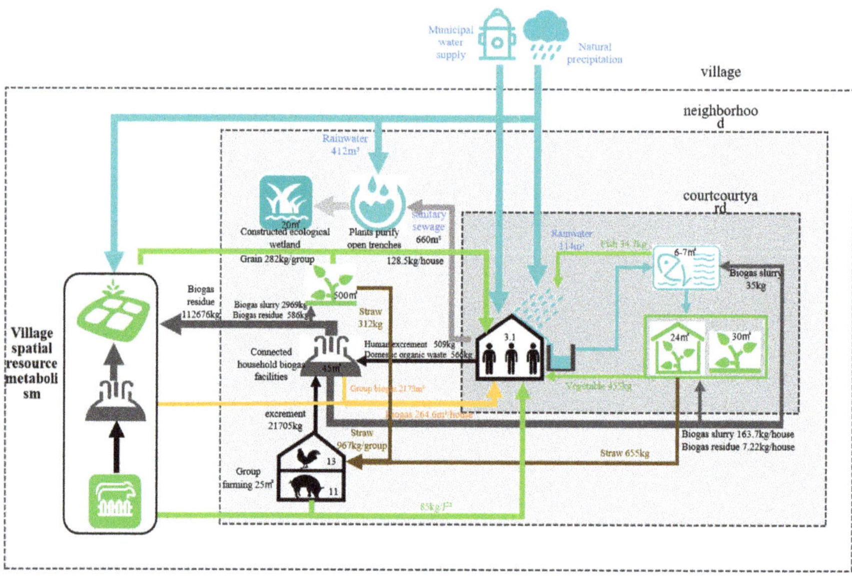

Figure 17. Resource metabolism modelling for neighborhoods.

(3) Whole Village

The metabolic accounting of village resources involved setting up a centralized breeding plant to raise 17 cows and 186 sheep, generating 134,228 kg of manure. A centralized breeding plant for 17 cows and 186 sheep generated 134,228 kg of manure. A centralized biogas station in the breeding space produced 8054 m^3 of biogas, 27,545 kg of biogas residue, and 18,363 kg of biogas slurry. The biogas was distributed to 170 households, each receiving 47.4 m^3, along with 217.3 m^3/household from group biogas plants, totaling 264.7 m^3/household. Each household received 264.7 m^3 of biogas, covering 83.5% of the annual cooking biogas consumption of 317 m^3 per household. The biogas slurry was primarily used for grain cultivation on the village's peripheral arable land. Combining 18,363 kg of biogas residue and 27,545 kg of biogas slurry from the central biogas station with 122,638 kg of biogas residue from 17 clusters provided 140,221 kg of biogas slurry. This mixture could fertilize 41 hectares of grain fields, producing 17,056 kg of grains and 15,416 kg of straw annually. The combined grain output from village farmland and group planting, totaling

21,850 kg, fell short of the annual food demand of 89,252 kg for 530 villagers. Additionally, the straw production was insufficient to meet the feed requirements for cattle and sheep farming (Table 8, Figure 18).

Table 8. Resource metabolism status and metabolic modelling at the village spatial level (530 people).

Metabolic State	Resource Type	Indicator (yr)	Clarification
Resource input	Digestate	112,676 kg	Volume of residual digestate from neighborhoods
	Livestock breeding feed (cattle and sheep)	312,354 kg	The amount of external straw feed still needed to be supplied
Internal metabolism (biology)	Grain stalk	15,416 kg	Cattle and sheep breeding feed
	Animal manure	134,228 kg	Biogas preparation at biogas plants
	Digestate	18,363 kg	Nutrient requirements for supplying cereals for arable cultivation
	Digestate	27,545 kg	
Resource export	Cereal cultivation	17,056 kg	Distribution to courtyards
	Livestock meat (cattle and sheep)	706 kg	
	Landfill gas	8054 m^3	

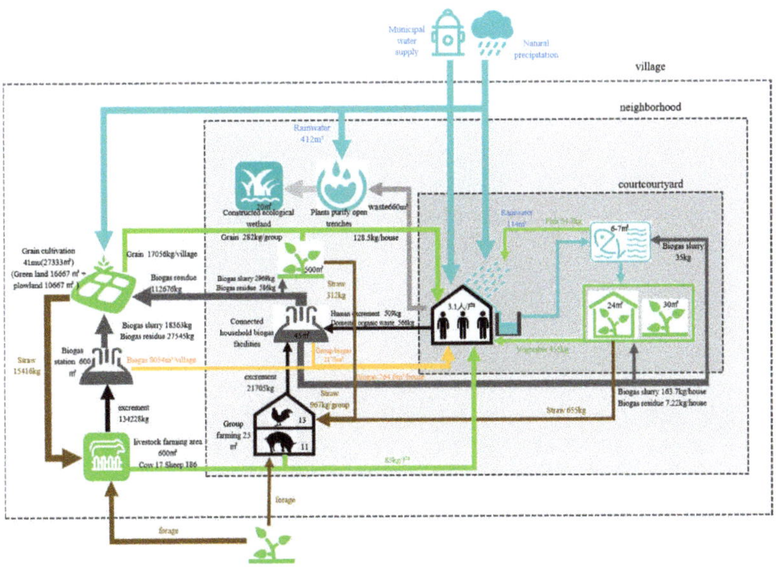

Figure 18. Resource metabolism modelling in the village.

Space allocation for resource metabolism facilities. The breeding area for 17 cows and 186 sheep covered approximately 600 m^2. Adjacent to it, the centralized biogas station, with an annual production capacity of 8054 m^3 of biogas, also occupied 600 m^2. For agricultural cultivation, there were 41 mu of land designated for grains, including 17,000 m^2 of public green space within the village. An additional 16 mu of cultivated land was provided, bringing the total cultivated area of the village to 622 mu.

Analyzing the metabolic flow of various resources at different scales and calculating the self-sufficiency index (SSI) based on the ratio of recycled resources to minimized demand revealed that, under ideal conditions, with 100% self-consumption of organic wastes, the self-sufficiency rate for vegetable cultivation at the yard scale of village residences was 105%, and for grain cultivation at the neighborhood and village scales, 24.5%. The self-sufficiency rate for straw from fruit, vegetable, and grain cultivation to meet livestock feed demand was 7.3%, and the self-sufficiency rate for cooking energy (biogas) was 83.5% (Table 9, Figure 19). The village had a remaining land area of 622 mu.

Table 9. Statistics on the self-sufficiency of major resources under the situation of organic waste self-consumption.

Resource Type	Quantity Demanded	Supply	Self-Sufficiency Index (SSI)	Metabolic Scale
Landfill gas	317 m^3/household	264.7 m^3/household	83.5%	Neighborhood + village
Fruits	434 kg/household	455 kg/household	105%	Courtyard
Grains	525 kg/household	128.5 kg/household	24.5%	Neighborhood + village
Stalk	465,288 kg/village	33,855 kg/village	7.3%	Courtyard + neighborhood + village

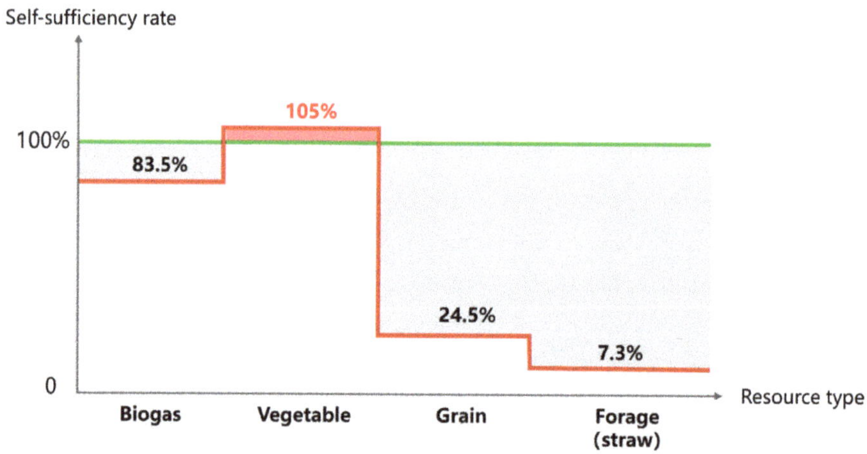

Figure 19. Self-sufficiency rates.

5.4.3. Resource Metabolism Spatial Modelling

The quantitative flow metabolism model and corresponding spatial occupation indices of the main resources for human life and farming production at the three scales—the courtyard, neighborhood, and village—are related from the bottom up. Main resource metabolism elements were allocated reasonably and effectively at each scale, in conjunction with the allocation of basic spatial indices and setting of morphological elements. This approach better adapted to and reflected the spatial layout model of rural settlements under the objective of "organic waste self-consumption". The spatial layout model of rural settlements can better adapt to and reflect the goal of "self-consumption of organic waste".

(1) Spatial Modelling of Courtyards

At the courtyard scale, based on the quantitative measurement of the resource metabolism model, the location and area of various resource metabolism and productive spaces were reasonably allocated (Figure 16). On a 200 m^2 residential base, excluding the village house area of about 90–100 m^2 and courtyard traffic, the remaining 40–50% open space could be used for agricultural cultivation. The roof and ground of a one-story village house provided about 54 m^2 for planting, with 24 m^2 of the roof suitable for a small greenhouse. Vegetable cultivation could assist in heat transfer to neighboring rooms, raising room temperatures in winter. Depending on vegetable yield targets, the courtyard' open vegetable garden could be allocated different areas and locations for resource metabolism and productive spaces (Figures 20 and 21). Depending on vegetable yield targets, the courtyard open vegetable garden could be converted into a small vegetable greenhouse. Pools of 6–7 m^2 near the courtyard wall could collect precipitation, breed fish, and form an aquaponics system with the planting gallery. Excess water, along with domestic sewage (gray water), could be discharged into the plant purification ditch along the external road, ultimately transferring to the group's small-scale artificial ecological purification wetland system.

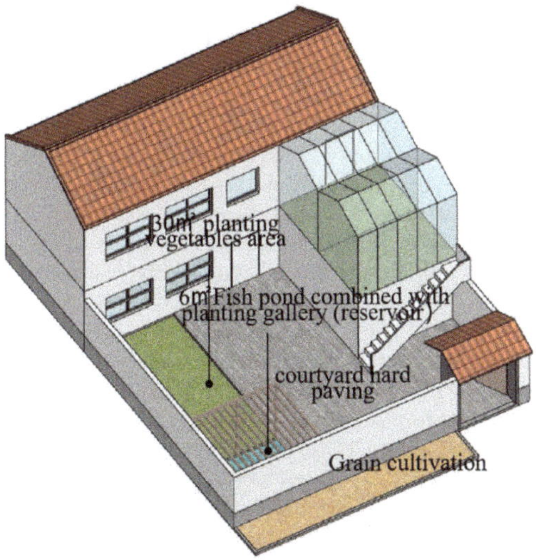

Figure 20. Spatial model of the courtyard.

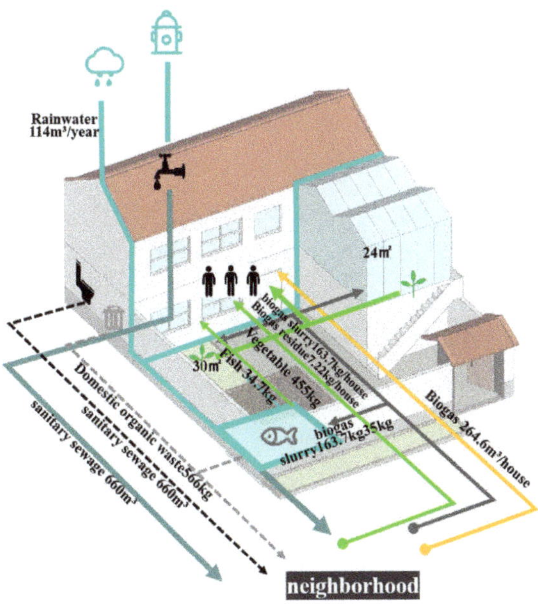

Figure 21. Metabolic spatial model of the courtyard.

(2) Spatial Modelling of Neighborhoods

The neighborhood comprised 10 courtyards responsible for the input, consumption, and distribution of most organic waste resources. Limited open space sites, often unused or abandoned areas, were utilized to integrate various resource-producing and -utilizing technologies and facilities, such as agricultural cultivation spaces, artificial ecological water purification wetlands, and small-scale household biogas digesters. The metabolic space model of the neighborhood was constructed based on the complete recycling and consumption of all types of living and production organic waste by 10 households. The location and area of various facilities, including biogas digesters, were reasonably configured and measured.

Based on the total amount of organic waste at the group scale, it was estimated that a continuous household biogas digester (underground) with an annual biogas production capacity of 2173 m^3 required 45 m^2 of floor space. It is advisable to establish livestock breeding and small-scale greenhouse spaces above it. In the group, 11 pigs and 13 meat birds were raised in a vertical layout. The breeding space was determined to be 25 m^2 based on the number of livestock. The remaining 20 m^2 served as a neighborhood-shared greenhouse. The complementary heat effect of the resource metabolism facilities ensures the underground biogas digester maintains a normal vapor production temperature in winter, while the waste heat from the anaerobic fermentation process insulates the livestock breeding space above. From estimating the artificial ecological wetland area, 10 households in the village produced a total of 660 m^3 of domestic sewage per year. Approximately 580 m^2 of roads and other hard surfaces in the group could collect 412 m^3 of rainwater, totaling 1072 m^3 of gray water. Based on a calculation of 3 m^3/day, the small-scale artificial ecological wetland occupied about 20 m^2 (including 10 m^2 of regulating ponds and 10 m^2 of stabilized ponds combined with lotus root planting, with a water bed depth of 0.7 m). The remaining block vegetable land in front of the house was 140 m^2, and there was 366 m^2 of centralized planting land in the public site, totaling about 500 m^2 mainly used for planting grains. Additionally, a fruit and vegetable planting corridor of about 290 m^2 could be established above the road (Figures 22 and 23). The 10 courtyards in the neighborhood could be flexibly deployed and combined with the open space, integrating suitable resources and metabolic technology facilities to form various group layouts such as enclosed and row types (Figure 24).

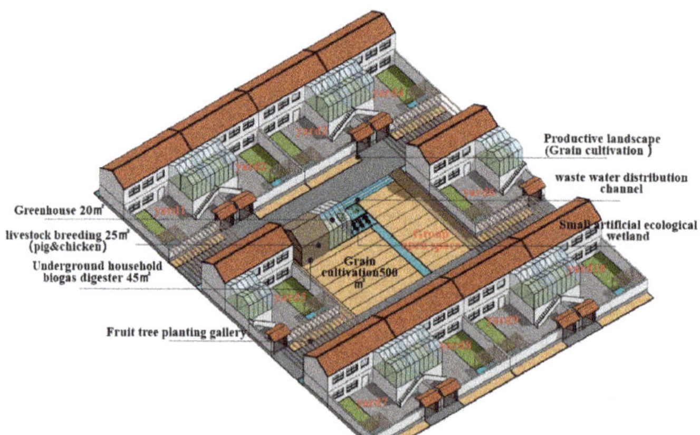

Figure 22. Layout of spatial facilities.

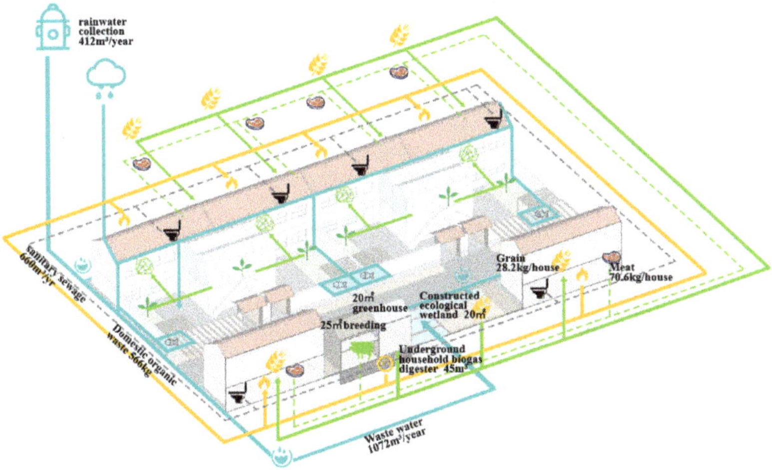

Figure 23. Schematic resource metabolism for the neighborhood spatial model.

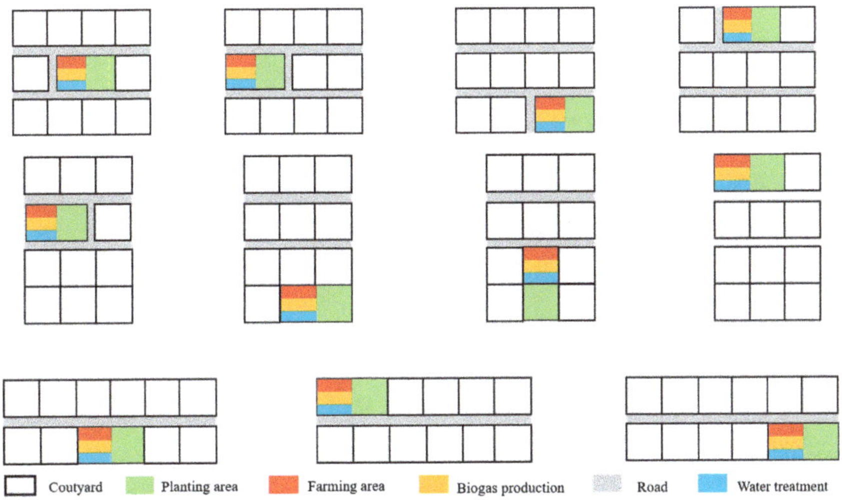

Figure 24. Neighborhood layout.

(3) Village Spatial Modelling

The village, housing 530 residents, covered an area of 766 mu (509,575 m^2). The living area was centralized into 17 neighborhoods, divided by a cross-shaped main road. These clusters were connected through streets and lanes. Land for public buildings (such as the village committee, commerce, health stations, and other public facilities) and green areas were concentrated in the village center. Public green spaces in the village were transformed into productive areas for grain cultivation to meet agricultural production needs and facilitate the consumption of organic waste. The primary resource metabolism facilities were centrally located downwind of the prevailing winds to ensure the efficient metabolism of organic waste and water resources. These facilities were separated from residential areas by green belts to minimize their impact on the living environment (Figure 25).

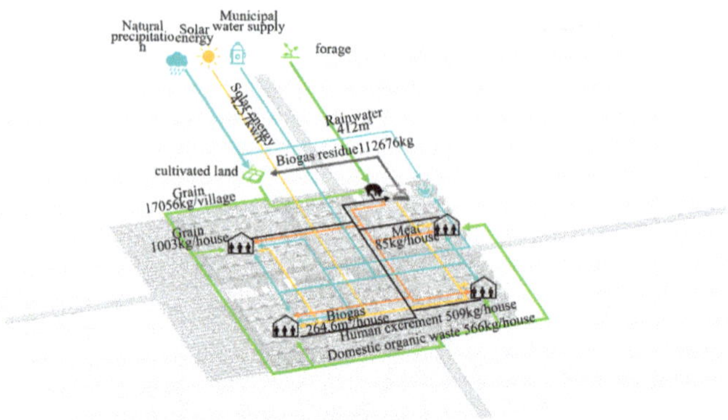

Figure 25. Spatial model of the spatial metabolism in the village area.

6. Conclusions

The resource metabolism modelling process under organic waste self-consumption, utilizing appropriate technologies and integrating spatial elements at different scales, achieved full cyclical metabolism of human and animal organic waste in rural areas. This approach saved significant arable land on the periphery of the village's built-up area. However, it is challenging to achieve self-sufficiency in food resources, necessitating the import of additional nutrients to meet food production needs. The neighborhood scale is crucial in transitioning resource flow and facilitating local transformation. By utilizing public land for livestock breeding, the cultivation of fruits, vegetables, or grains, and water purification and reuse, it is possible to efficiently transform various organic waste resources from village yards. This significantly reduces and decentralizes the burden of constructing and operating resource metabolism facilities at the village scale. This approach also minimizes the distance and process losses associated with resource transmission and distribution. At the courtyard and neighborhood scales, various unused spatial resources are effectively utilized to establish facilities for organic waste resource utilization, achieving the efficient integration of production, living, and ecological spatial elements. While the complete recycling of organic waste from the village population (530 people) and their livestock, meeting nutrient needs for basic food cultivation (especially grains) and cooking energy, remained challenging, modelling the complete consumption of organic waste helped for understanding the dynamic coupling between organic waste consumption and the available space for agricultural support. Modelling organic waste adequacy helps elucidate the dynamic coupling between organic waste consumption and the spatial requirements for agricultural support.

Author Contributions: Conceptualization, M.L. and Z.Z.; methodology, M.L.; software, M.L.; validation, M.L. and Z.Z.; formal analysis, M.L.; investigation, J.Z.; resources, J.Z.; data curation, C.L.; writing—original draft preparation, Z.Z.; writing—review and editing, M.L.; visualization, M.L.; supervision, J.Z.; project administration, C.L.; funding acquisition, Z.Z. All authors have read and agreed to the published version of the manuscript.

Funding: This work was supported by the Natural Science Foundation of Shandong Province of China (General Program) (Grant No. ZR2023ME229) and China. & Shandong Provincial Key R&D Program (Soft Science Project) (No. 2024RKY0604). and "the Case Repository for the Revitalisation and Adaptive Reuse of Industrial Heritage" (Grant No. SDYAL2022157) funded by the Graduate Teaching Case Repository for Professional Degrees in Shandong Province, China.

Data Availability Statement: The data that support the findings of this study are available from the corresponding author [LIU Meng] upon reasonable request.

Conflicts of Interest: ZHOU Zhongkai reports having received financial support provided by the Shandong Provincial Department of Science and Technology, the Natural Science Foundation of Shandong Province of China, and the Graduate Teaching Case Repository for Professional Degrees in Shandong Province. If there are other authors, they declare that they have no known competing financial interests or personal relationships that could have appeared to have influenced the work reported in this paper.

References

1. Wolman, A. The metabolism of cities. *Sci. Am.* **1965**, *213*, 178–193. [CrossRef]
2. Rappaport, R.A. The flow of energy in an agricultural society. *Sci. Am.* **1971**, *224*, 117–133.
3. Ayres, R.U. Industrial metabolism: Theory and policy. In *The Greening of Industrial Ecosystems*; Allenby, B.R., Richards, D.J., Eds.; National Academy Press: Washington, DC, USA, 1994.
4. Alberti, M. Measuring urban sustainability. *Environ. Impact Assess. Rev.* **1996**, *16*, 381424. [CrossRef]
5. Newman, P.W.G. Sustainability and cities: Extending the metabolism modelf1l. *Land Use Urban Plan.* **1999**, *44*, 219–226. [CrossRef]
6. Agudelo-Vera, C.M.; Mels, A.R.; Keesman, K.J.; Rijnaarts, H.H. Resource management as a key factor for sustainable urban planning. *J. Environ. Manag.* **2011**, *92*, 2295–2303. [CrossRef]
7. Girardet, H. The Metabolism of Cities. *Sci. Am.* **1990**, *213*, 179–190.
8. Kennedy, C.; Pincetl, S.; Bunje, P. The study of urban metabolism and its applications to urban planning and design. *Environ. Pollut.* **2011**, *159*, 1965–1973. [CrossRef]
9. Roggema, R. *TransFEWmation: Towards Design-Led Food-Energy-Water Systems for Future Urbanization*; Springer: Cham, Switzerland, 2021.
10. Tillie, N.; Van Den Dobbelsteen, A.; Doepel, D.; Joubert, M.; De Jager, W.; Mayenburg, D. Towards CO_2 Neutral Urban Planning: Presenting the Rotterdam Energy Approach and Planning (REAP). *J. Green Build.* **2009**, *4*, 103–112. [CrossRef]
11. Jan, J.; Nelson, N.; Goosens, F. Recyclicity: A Toolbox for Resource-Based Design. *Archit. Des.* **2011**, *81*, 54–61.
12. Zhao, Q. *A Primer on Urban Efficiency—Research Practices Based on Metabolic Efficiency in Urban Settlements*; Science Press: Beijing, China, 2017; pp. 237–253.
13. Yue, X.; Zhang, Y. The research about concept evolution and development process of ecovillage abroad. *Archit. J.* **2011**, 123–129. Available online: https://xueshu.baidu.com/usercenter/paper/show?paperid=207dc80508da27ffd0c6491915f000a2&site=xueshu_se (accessed on 10 October 2024).
14. Bian, Y. *Reunion Eco-Agricultural Systems*; China Environmental Science Press: Beijing, China, 1988.
15. Weng, B.; Huang, Q.; Chen, J. Construction of Eco-Village-New Development of Sustainable Agriculture. *Yunnan Environ. Sci.* **2000**, *19*, 99–103.
16. Chen, Y.; Yang, Y. Comments and prospects of eco-village construction in China. *North. Environ.* **2011**, *23*, 71–74.
17. Zhao, J.; Zhang, Y. Contemporary Western Experiment on Integration of Urban Agriculture and Urban Space. *New Archit.* **2012**, 27–31. Available online: https://xueshu.baidu.com/usercenter/paper/show?paperid=46b7ec95aee4e9a5e3fb5d696e598357&site=xueshu_se (accessed on 10 October 2024).
18. Baccini, P.; Brunner, P.H. *Metabolism of the Anthroposphere: Analysis, Evaluation, Design*; The MIT Press: Cambridge, MA, USA, 2012.
19. Zhou, Z.; Yang, X.; Zhao, J. The Rural Settlements Design and Practice Under the Resource Metabolism Theory and Its Influence. *Architect* **2021**, 75–83. Available online: http://thearchitect.cabp.com.cn/ch/reader/view_abstract.aspx?file_no=20210508 (accessed on 10 October 2024).
20. Chen, X.; Wu, L.; Dang, Y. The Inherent Mechanism and Enlightenment of Production-Living-ecological Space in Traditional Human Settlement Environment. *Urban Dev. Stud.* **2022**, *29*, 6–11.
21. Chen, X. Spatial Characters and Restructuring Keynotes of the Rural Area in China. *Planners* **2007**, 79–82. Available online: https://xueshu.baidu.com/usercenter/paper/show?paperid=3c439077ab54f58e8485fa1a78cb50dc&site=xueshu_se (accessed on 10 October 2024).
22. Korhonen, J. Two paths to industrial ecology: Applying the product-based and geographical approaches. *J. Environ. Plan. Manag.* **2002**, *45*, 39–57. [CrossRef]
23. Zhou, Z.; Yang, R.; Hao, X.; Zhao, J. Research on the Optimal Design Path and Strategy of Rural Space from the Metabolism Perspective: A Case Study of Xishan Village, Zibo City. *Mod. Urban Res.* **2023**, *8*, 50–58.
24. Forman, R.T. Land mosaics: The ecology of landscapes and regions 1995. In *The Ecological Design and Planning Reader*; Island Press: Washington, DC, USA, 2014; pp. 217–234.
25. Zhou, Z.; Chen, Q.; Zhang, H.; Wu, W. Study on the "Four-in-One" eco-agricultural model. *Renew. Energy Resour.* **2007**, 93–96. Available online: https://xueshu.baidu.com/usercenter/paper/show?paperid=297700d6935bae3ff31930b20406d2a9&site=xueshu_se (accessed on 10 October 2024).
26. Fu, Y. Pleasure of Water Gathered: Restoring the Public Space in Villages Based on Ecological Demonstration on the Design of an Ecological Rainwater Garden in Lianma Village, Guangzhou City. *Archit. J.* **2016**, 101–103. Available online: https://www.gooood.cn/pleasure-of-water-gathered.htm (accessed on 10 October 2024).
27. John, L.; Joshua, B.; Zhang, J. Rural Urban Framework (RUF)/The University of Hong Kong; Jintai Village Reconstruction, Bazhong, China 2017. *World Archit.* **2018**, *36*. Available online: https://www.archdaily.com/882714/jintai-village-reconstruction-rural-urban-framework (accessed on 10 October 2024).

28. Liu, C.; Zhang, Y.; Zhao, J. Research on urban agro-integrated community based on circular metabolism. *City Plan. Rev.* **2018**, *42*, 52–59.
29. Meng, K. Construction of Closed-Loop Model of Material Energy in Tianjin Eco-Village Based on System Dynamics and Research on Spatial Layout. Master's Thesis, Hebei University of Technology, Tianjin, China, 2022. [CrossRef]
30. Yiu, E.C.Y.; Castro Campos, B. *A Novel Planning Model for Approaching a Zero-Food and Zero-Waste Community*; Social Science Electronic Publishing: New York, NY, USA, 2014.
31. Leduc, W.R.W.A.; Rovers, R. Urban Tissue: The representation of the Urban Energy Potential. In Proceedings of the PLEA 2008-25th Conference on Passive and Low Energy Architecture, Dublin, Ireland, 22–24 October 2008.
32. Takeuchi, K.; Namiki, Y.; Tanaka, H. Designing eco-villages for revitalizing Japanese rural area. *Ecol. Eng.* **1998**, *11*, 177–197. [CrossRef]
33. Song, W. Study on the Spatial Form of Traditional Villages in Shandong. Ph.D. Thesis, Dalian University of Technology, Dalian, China, 2021.
34. Shandong Provincial Department of Water Resources. *Shandong Provincial Water Conservancy Yearbook 2012*; Shandong Provincial Map Publishing House: Jinan, China, 2013.
35. Sheng, S.; Wang, Y. Rural Ecological Restoration Strategies in Plain Agricultural Areas Based on the Identification of Ecological Characteristics and Issues through Clustering—A Case Study of Heishan County, Liaoning Province. *Chin. Landsc. Archit.* **2022**, *38*, 20–25.
36. Shandong Provincial Department of Water Resources. *The 13th Five-Year Plan for Water Conservancy Development in Shandong Province Shandong*; Shandong Provincial Department of Water Resources: Jinan, China, 2020.
37. Liu, C.; Zhao, J. *Integrated Community in Cities*; China Architecture & Building Press: Beijing, China, 2016.
38. People's Republic of China Bureau of Statistics. *China Rural Statistical Yearbook*; China Statistics Press: Beijing, China, 2021.
39. Shandong Provincial Bureau of Statistics. *Shandong Statistical Yearbook*; China Statistics Press: Beijing, China, 2021.
40. Shandong Provincial Bureau of Statistics. *Statistical Communiqué on the National Economic and Social Development of Shandong Province in 2022*; China Statistics Press: Beijing, China, 2021.
41. Zhu, J.; Huang, Z.; Zou, C.; Jiang, Y.; Wu, Y. Centralized (Joint Households) Gas Supply Mode of Rural Small-scale Biogas Projects in Ji'an County. *China Biogas* **2008**, *26*, 34–36.
42. A Pattern Language. In *Christopher Alexander*; China Architecture & Building Press: Beijing, China, 1989.
43. Wang, Z.; Fan, L.; Chen, Z. Study on the Pattern of "Eco-habitation" in New Countryside: Case Study of Countryside in Southern Yangtze River Areas of China. *Archit. J.* **2011**, 22–26. Available online: https://xueshu.baidu.com/usercenter/paper/show?paperid=93bf49a58e7dcb0e0133d1e34999f6c9&site=xueshu_se (accessed on 10 October 2024).

buildings

MDPI

Article

Examining the Impact of the Built Environment on Multidimensional Urban Vitality: Using Milk Tea Shops and Coffee Shops as New Indicators of Urban Vitality

Ziqi Xu [1], Jiang Chang [1,*], Fangyu Cheng [2], Xiaoyi Liu [1], Tianning Yao [1], Kuntao Hu [1] and Jingyu Sun [3]

[1] School of Architecture and Design, China University of Mining and Technology, Xuzhou 221116, China; ts22190042p31@cumt.edu.cn (Z.X.); ts23190006a31@cumt.edu.cn (X.L.); tianning_yao@cumt.edu.cn (T.Y.); ts22190026p31@cumt.edu.cn (K.H.)

[2] School of Architecture and Design, Harbin Institute of Technology, Harbin 150001, China; fcheng@hit.edu.cn

[3] School of Materials Science and Engineeing, Shanghai University of Engineering Science, Shanghai 201620, China; angjusun@outlook.com

* Correspondence: changjiang@cumt.edu.cn

Abstract: Urban vitality is a critical driver of sustainable urban development, significantly contributing to the enhancement of human well-being. A thorough and multidimensional comprehension of urban vitality is essential for shaping future urban planning and policy-making. This study, focused on Chengdu, proposes a framework for assessing various dimensions of UV through the distribution of milk tea and coffee shops. Using random forest and multi-scale geographically weighted regression models, this study investigates the factors influencing urban vitality from both mathematical thresholds and spatial heterogeneity, and develops spatial maps of future vitality to inform targeted urban strategies. The results show that (1) the milk tea index is effective in capturing population vitality, while the coffee index is more closely associated with economic vitality and urban renewal; (2) office buildings (13.46%) and commercial complexes (13.70%) have the most significant impact on both economic and population vitality, while the importance of transportation factors has notably decreased; (3) the influence of these factors demonstrates spatial heterogeneity and nonlinear relationships, with subway station density of 0.5–0.8 stations per kilometer being optimal for stimulating both types of vitality. The minimum threshold for economic vitality in a given unit is a housing price exceeding 6000 RMB/m^2; (4) the future vitality map suggests that urban planners should pay greater attention to non-central districts with high development potential. Moreover, spontaneous social interactions and consumer behaviors stimulated by various shops are critical components of urban vitality. In designing the physical environment and urban spatial forms, special attention should be given to enhancing the attractiveness of physical spaces and their capacity to accommodate social interaction.

Keywords: urban vitality; milk tea and coffee shops; random forest; MGWR model; nonlinear relationships; social interactions

1. Introduction

The concept of "urban vitality (UV)" originated in the 1960s and has undergone significant evolution and enrichment through the examination of urban morphology and social interactions by Western scholars. Research across diverse disciplines, including geography, sociology, and urban management, has consistently identified UV as a fundamental metric for evaluating urban quality of life [1]. Urban spaces with high vitality are widely regarded as optimal environments for residential, occupational, and recreational purposes, as they foster social cohesion and contribute significantly to the overall well-being of inhabitants. Furthermore, such spaces are instrumental in attracting human capital, stimulating urban innovation, and mobilizing social investment, thereby enhancing urban competitiveness.

Further inquiry into UV must address three core questions: what constitutes UV, how can it be effectively represented, and what factors influence it? Jacobs, who introduced this concept to urban studies, posited that UV emerges from the diversity of life generated by the interweaving of citizen activities and living spaces, with "street life over a 24 h period" being central to the concept of UV [2]. Lynch, from the perspective of spatial morphology, emphasized that the essence of this concept lies in the capacity of settlement morphology to support and accommodate human activities, ecological requirements, and life functions [3]. Meanwhile, Maas deconstructed UV into three essential components: the continuous presence of people in public spaces, their activities, and the environments in which these activities occur [4]. These theoretical perspectives underscore the intricate and complex interactions between human activities and spatial morphology. Kim attempted to conceptualize UV from a more macro-level perspective, extending its theoretical framework to encompass economic, social, and cultural dimensions based on urban sociological theories [5]. The inherent complexity of cities and populations has resulted in an absence of consensus regarding a unified definition of UV; however, the notion that "UV is a multidimensional concept encompassing population, social, economic, and spatial aspects" has gradually gained widespread acceptance [6]. Therefore, this study will primarily focus on the population and economic dimensions, with particular emphasis on how built environment elements facilitate the synergy between dense pedestrian activities and economic consumption.

UV consists primarily of two key aspects: crowd aggregation and place attractiveness. The precise quantification of UV plays a crucial role in objectively assessing the quality of urban development [7]. Traditional methods for measuring human activity, such as field observations, often fell short in effectively capturing large-scale patterns of human movement within urban environments [8,9]. With the development of information technology, large-scale, high spatiotemporal precision "volunteered geographical information" (such as mobile phone signaling data, social media data, and Wi-Fi data) has gradually become a mainstream method for dynamically monitoring and accurately quantifying UV. The advent of large-scale, high-resolution "volunteered geographical information (VGI)"—encompassing mobile location data [10], cell phone signal data [11], social media data [12]—has provided a prominent tool for dynamically monitoring and accurately quantifying UV. Despite these advancements, challenges related to data accessibility have somewhat limited its broader application. Numerous studies have also introduced physical urban elements, such as road intersection density, points of interest (POI) diversity, land-use heterogeneity, and residential price per square meter, as indirect indicators of UV [13–15]. In recent years, night-time light data (NTL), which integrate urban spatial morphology with the intensity of human socio-economic activities, have been widely employed to quantify UV and development, particularly showing strong performance in areas such as urban expansion, economic growth, and energy consumption [16–18]. Nonetheless, it is imperative to account for the intrinsic discrepancies in how different data types capture and represent UV [19]. For instance, university campuses may exhibit high densities of mobile signaling and social media check-ins, yet show relatively low POI distributions, which may lead to potential misinterpretations of vitality in such settings. Therefore, the implementation of a multidimensional analytical framework is essential to ensure a more comprehensive and accurate representation of UV, thereby allowing researchers to derive a deeper and more nuanced understanding of urban dynamics.

Small-scale food venues, such as coffee shops (COSs), beverage outlets, and fast-food restaurants, have recently been recognized as indirect indicators of UV. These venues not only attract crowds but also support diverse economic and social activities within urban streets and communities [20,21]. Their dependence on a stable customer base and sensitivity to environmental changes make them ideal for reflecting the real-time state of UV [22]. Studies conducted in various cities, including the United States [23], Berlin [24], Shenzhen [25], and Beijing [26], have confirmed this observation. COSs are typically located in densely populated and highly walkable areas [27,28], functioning as "third spaces" that provide platforms for social interaction and co-working [29], while also hosting significant economic activities. Quantitative research by Donner and Guzman has highlighted that communities with Starbucks outlets tend to demonstrate more pronounced economic and innovative development outcomes [30,31]. Glaeser et al. further identified a strong correlation between the presence of Starbucks and key indicators of UV, such as housing prices [32]. Shi et al. employed the "Starbucks Index" calculated based on the number of Starbucks outlets, as a proxy indicator for urban commercial activity and conducted a study that used COSs to represent urban economic vitality [33].

Milk tea, a traditional Chinese beverage [34], has seen rapid global expansion since its rise in popularity around the 2000s [35], with a market size reaching $2.46 billion by 2023 [36]. Despite the growing influence of coffee culture in China [37], consumption data show that tea-based beverages remain dominant [38,39]. With milk tea shops (MTSs) expanding at an annual rate of 15% [40], they have become a notable feature in China's urban landscape. However, most existing research has focused on aspects such as the health impacts of milk tea [35], consumer behavior [41], spatial design [42], and the brand management [43], while little attention has been paid to the relationship between MTSs and UV. It remains unclear whether MTSs, like COSs, can serve as "indicator locations" of UV, and if so, what aspects they might represent. This study, based on a systematic exploration of Chengdu, provides evidence on the potential of MTSs to serve as indicators of UV and explores the aspects they may represent.

In-depth investigation of the spatial drivers behind UV is crucial for formulating targeted policies to promote vitality. Human activities and socio-economic factors are widely recognized as fundamental influences on UV [44,45], while built environment elements—such as facility diversity [46], transportation accessibility [47], mixed land-use [48], urban texture [10], and street color [49]—have also been extensively confirmed to affect the intensity of UV. Adequate infrastructure and well-designed street facades often support more diverse human activities, fostering vibrant streets and communities. Traditional models, such as ordinary least squares (OLS) regression and geographically weighted regression (GWR), have made important contributions to understanding the relationship between the built environment and UV [50,51]. However, the complexity and diversity of urban spatial structures pose challenges for these models in addressing spatial heterogeneity and nonlinear relationships between variables [52]. The fixed bandwidth analysis in GWR models may not fully capture spatial scale differences, and the assumed linear relationship between the built environment and UV has been questioned in the context of rapid urban development [53]. Machine learning (ML) models, which are adept at identifying nonlinear relationships, and multi-scale geographically weighted regression models (MGWR), which account for variations in spatial scales, present new opportunities for advancing our understanding of this relationship [54,55].

In light of the above discussion, this study explores the utility of the "Milk Tea Index (MTI)" and the "Coffee Index (COI)" in representing UV, using Chengdu as a case study. We employ the MGWR model and random forest (RF) to investigate the relationship between built environment factors and UV. Finally, we generate future vitality maps based on the findings to expand the deeper application of these indices. Therefore, this study aims to (1) utilize coffee and milk tea shops to measure different dimensions of UV, compare their similarities and differences, and provide a new methodological reference for the quantitative assessment of UV; (2) explore the relationship between built environment characteristics and UV represented by the MTI and COI while accounting for nonlinear relationships and spatial heterogeneity; (3) propose region-specific strategies to enhance UV based on the findings.

2. Materials and Methods

2.1. Study Area

This study examines 12 administrative districts in Chengdu, each exhibiting distinct levels of development (Figure 1). Historically, Chengdu has been recognized as a highly livable city with a strong foundation in commerce and trade. Today, it ranks as one of the most economically developed and commercially dynamic regions in western China. Chengdu consistently appears at the top of the "Forbes China Urban Consumption Vitality" index, with its permanent population increasing by nearly 6 million over the past decade. In 2023, Chengdu's food and beverage retail sales accounted for 15.88% of the city's total, significantly surpassing other major cities such as Beijing and Shanghai, positioning Chengdu as the leading city in China for dining expenditures. In the same year, the number of milk tea outlets in Chengdu ranked third nationwide, behind Guangzhou and Shenzhen, while its number of independent coffee shops was second only to Shanghai. Chengdu is emerging as a "beverage capital" of western China [56].

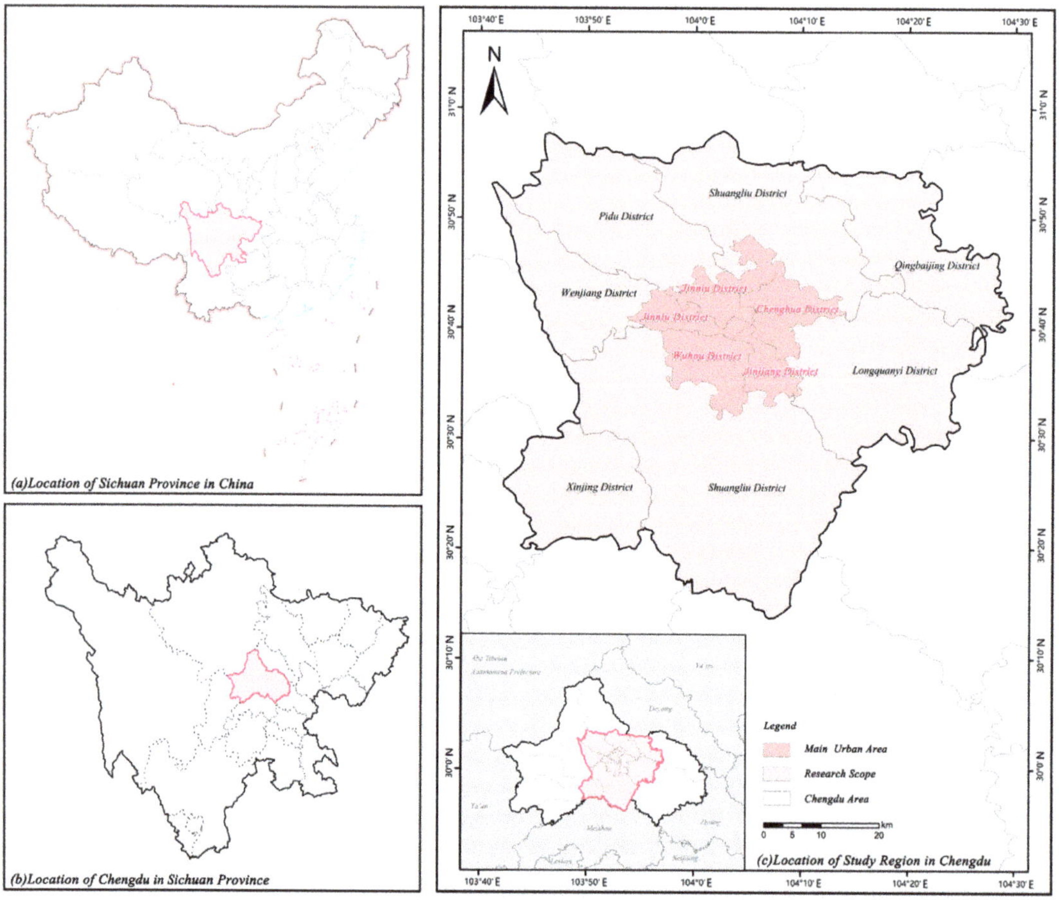

Figure 1. Overview map of the study area((**a**) Location of Sichuan Province in China. (**b**) Location of Chengdu in Sichuan Province. (**c**) The specific composition of the study area).

2.2. Data Sources and Processing

Considering the synthesis of existing research, this paper employs three categories of data: remote sensing imagery, open-source big data, and traditional statistical data. The data type, data source, and acquisition time are shown in Table 1.